Climatology: An Atmospheric Science

Climatology: An Atmospheric Science

Edited by
Braxton Stewart

Larsen & Keller
www.larsen-keller.com

Climatology: An Atmospheric Science
Edited by Braxton Stewart
ISBN: 978-1-63549-069-5 (Hardback)

© 2017 Larsen & Keller

⊟ Larsen & Keller

Published by Larsen and Keller Education,
5 Penn Plaza,
19th Floor,
New York, NY 10001, USA

Cataloging-in-Publication Data

Climatology : an atmospheric science / edited by Braxton Stewart.
         p. cm.
Includes bibliographical references and index.
ISBN 978-1-63549-069-5
1. Climatology. 2. Climatic changes. 3. Meteorology. 4. Atmosphere.
I. Stewart, Braxton.
QC861.3 .C55 2017
551.5--dc23

The publisher's policy is to use permanent paper from mills that operate a sustainable forestry policy. Furthermore, the publisher ensures that the text paper and cover boards used have met acceptable environmental accreditation standards.

Printed and bound in the United States of America.

For more information regarding Larsen and Keller Education and its products, please visit the publisher's website www.larsen-keller.com

Table of Contents

Preface

As a subfield of physical geography and part of atmospheric sciences, climatology refers to the study of the climate or the weather condition of a place at or over a particular time. It includes fields like biogeochemistry and oceanography. It is applied to forecast weather and study the changes in climate and analyze the results for understanding the effects of phenomena like greenhouse effect and pollution have on climate. This book is a compilation of chapters that discuss the most vital and fundamental concepts in the field of climatology. The topics covered in this extensive book deal with the core areas of this subject. Coherent flow of topics, student-friendly language and extensive use of examples make this book an invaluable source of knowledge.

To facilitate a deeper understanding of the contents of this book a short introduction of every chapter is written below:

Chapter 1- Climatology studies climate patterns occurring over a defined period of time. This can be records of the past or present and future weather forecasting. Climatology is an emerging field of study; the following chapter will not only provide an overview, it will also delve deep into the variegated topics related to it.

Chapter 2- The earth's climate is dependent on many factors such as volcanic eruptions and asteroid impacts, frequent during the early stages of the evolution of earth, to rising temperature levels caused by the emission of greenhouse gases. This chapter seeks to explain the recent additions to climate change records as well as instances of drastic climate change in the ancient past.

Chapter 3- Climate is dependent on location and this demarcates the intensity with which a region experiences its climate. Climate defines the flora and fauna of a region, from tropical rainforests to desert regions. Climate is formed over a significant period of time. This chapter elucidates the crucial theories and principles of climatology.

Chapter 4- Climatology has gained popularity in recent years due to its ability to extract climate records from the past. Climatologists are able to compare as well as predict climate patterns, especially natural disasters and storms through the creation of climate models. This chapter is a compilation of the various branches of climatology that form an integral part of the broader subject matter.

Chapter 5- Greenhouse gas emissions and the use of aerosols have led to dangerous levels of increase of surface temperature levels on earth. At the same time, climate change can be defined as a great climatic event that influences more than one region. The earth has gone through many climate changes, as has been explained in this chapter.

Chapter 6- By recording climate patterns over a large period of time, it is possible to create climate models that can predict future climatic patterns such as tidal movements, precipitation and levels of atmospheric pressure. Such models can also help in meteorology and weather sciences. Climatology is best understood in confluence with the major topics listed in the following chapter.

Chapter 7- Climate changes along periods that recur over and over and these repetitive "oscillatory" patterns influence the habitation and growth of vegetation, animals and human beings. This chapter concentrates on climate patterns found near the different continents as well as on classifications such as climate state that has been in use in climatology studies.

Chapter 8- With the failure to reduce greenhouse gas emissions, scientists have turned to the study of present and possible future disasters as well as extreme weather that certain regions of the world has encountered. This chapter is an overview of the subject matter incorporating all the major aspects of global warming.

Finally, I would like to thank the entire team involved since the inception of this book for their valuable time and contribution. This book would not have been possible without their efforts. I would also like to thank my friends and family for their constant support.

Editor

An Introduction to Climatology

Climatology studies climate patterns occurring over a defined period of time. This can be records of the past or present and future weather forecasting. Climatology is an emerging field of study; the following chapter will not only provide an overview, it will also delve deep into the variegated topics related to it.

Climatology or climate science is the study of climate, scientifically defined as weather conditions averaged over a period of time. This modern field of study is regarded as a branch of the atmospheric sciences and a subfield of physical geography, which is one of the Earth sciences. Climatology now includes aspects of oceanography and biogeochemistry. Basic knowledge of climate can be used within shorter term weather forecasting using analog techniques such as the El Niño–Southern Oscillation (ENSO), the Madden–Julian oscillation (MJO), the North Atlantic oscillation (NAO), the Northern Annular Mode (NAM) which is also known as the Arctic oscillation (AO), the Northern Pacific (NP) Index, the Pacific decadal oscillation (PDO), and the Interdecadal Pacific Oscillation (IPO). Climate models are used for a variety of purposes from study of the dynamics of the weather and climate system to projections of future climate.

History

Chinese scientist Shen Kuo (1031–1095) inferred that climates naturally shifted over an enormous span of time, after observing petrified bamboos found underground near Yanzhou (modern day Yan'an, Shaanxi province), a dry-climate area unsuitable for the growth of bamboo.

Early climate researchers include Edmund Halley, who published a map of the trade winds in 1686 after a voyage to the southern hemisphere. Benjamin Franklin (1706–1790) first mapped the course of the Gulf Stream for use in sending mail from the United States to Europe. Francis Galton (1822–1911) invented the term *anticyclone*. Helmut Landsberg (1906–1985) fostered the use of statistical analysis in climatology, which led to its evolution into a physical science.

Different Approaches

Climatology is approached in a variety of ways. Paleoclimatology seeks to reconstruct past climates by examining records such as ice cores and tree rings (dendroclimatology). Paleotempestology uses these same records to help determine hurricane frequency over millennia. The study of contemporary climates incorporates meteorological data accumulated over many years, such as records of rainfall, temperature and atmospheric composition. Knowledge of the atmosphere and its dynamics is also embodied in models, either statistical or mathematical, which help by integrating different observations and testing how they fit together. Modeling is used for understanding past, present and potential future climates. Historical climatology is the study of climate as related to human history and thus focuses only on the last few thousand years.

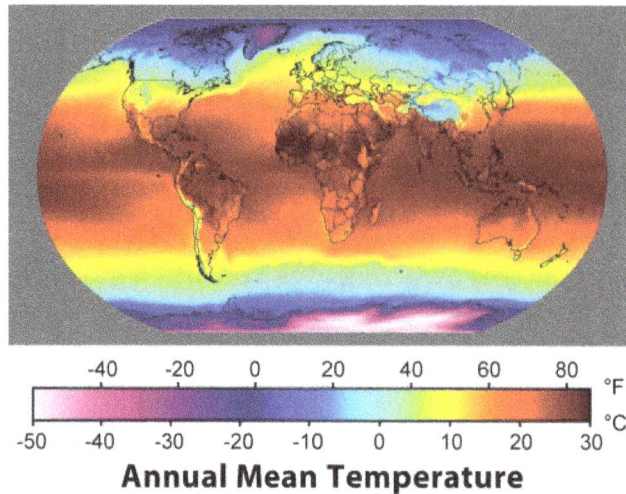

Annual Mean Temperature

Map of the average temperature over 30 years. Data sets formed from the long-term average of historical weather parameters are sometimes called a "climatology".

Climate research is made difficult by the large scale, long time periods, and complex processes which govern climate. Climate is governed by physical laws which can be expressed as differential equations. These equations are coupled and nonlinear, so that approximate solutions are obtained by using numerical methods to create global climate models. Climate is sometimes modeled as a stochastic process but this is generally accepted as an approximation to processes that are otherwise too complicated to analyze.

Indices

Scientists use climate indices based on several climate patterns (known as modes of variability) in their attempt to characterize and understand the various climate mechanisms that culminate in our daily weather. Much in the way the Dow Jones Industrial Average, which is based on the stock prices of 30 companies, is used to represent the fluctuations in the stock market as a whole, climate indices are used to represent the essential elements of climate. Climate indices are generally devised with the twin objectives of simplicity and completeness, and each index typically represents the status and timing of the climate factor it represents. By their very nature, indices are simple, and combine many details into a generalized, overall description of the atmosphere or ocean which can be used to characterize the factors which impact the global climate system.

El Niño–southern Oscillation

El Niño–Southern Oscillation (ENSO) is a global coupled ocean-atmosphere phenomenon. The Pacific ocean signatures, El Niño and La Niña are important temperature fluctuations in surface waters of the tropical Eastern Pacific Ocean. The name El Niño, from the Spanish for "the little boy", refers to the Christ child, because the phenomenon is usually noticed around Christmas time in the Pacific Ocean off the west coast of South America. La Niña means "the little girl". Their effect on climate in the subtropics and the tropics are profound. The atmospheric signature, the Southern Oscillation (SO) reflects the monthly or seasonal fluctuations in the air pressure difference between Tahiti and Darwin. The most recent occurrence of El Niño started in September 2006 and lasted until early 2007.

WARM EPISODE RELATIONSHIPS   DECEMBER - FEBRUARY

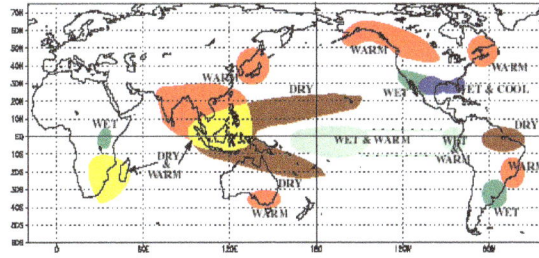

WARM EPISODE RELATIONSHIPS   JUNE - AUGUST

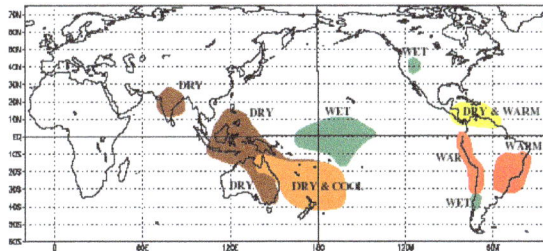

El Niño impacts

ENSO is a set of interacting parts of a single global system of coupled ocean-atmosphere climate fluctuations that come about as a consequence of oceanic and atmospheric circulation. ENSO is the most prominent known source of inter-annual variability in weather and climate around the world. The cycle occurs every two to seven years, with El Niño lasting nine months to two years within the longer term cycle, though not all areas globally are affected. ENSO has signatures in the Pacific, Atlantic and Indian Oceans.

COLD EPISODE RELATIONSHIPS   DECEMBER - FEBRUARY

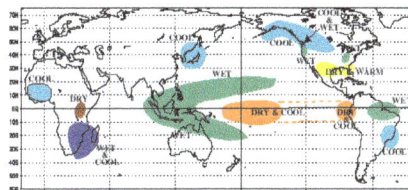

COLD EPISODE RELATIONSHIPS   JUNE - AUGUST

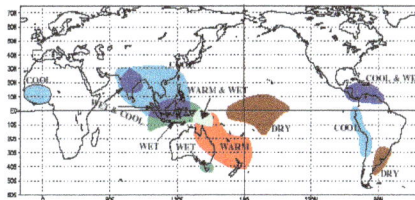

La Niña impacts

In the Pacific, during major warm events, El Niño warming extends over much of the tropical Pacific and becomes clearly linked to the SO intensity. While ENSO events are basically in phase between the Pacific and Indian Oceans, ENSO events in the Atlantic Ocean lag behind those in the Pacific by 12–18 months. Many of the countries most affected by ENSO events are developing countries within tropical sections of continents with economies that are largely dependent upon their agricultural and fishery sectors as a major source of food supply, employment, and foreign exchange. New capabilities to predict the onset of ENSO events in the three oceans can have global socio-economic impacts. While ENSO is a global and natural part of the Earth's climate, whether its intensity or frequency may change as a result of global warming is an important concern. Low-frequency variability has been evidenced: the quasi-decadal oscillation (QDO). Inter-decadal (ID) modulation of ENSO (from PDO or IPO) might exist. This could explain the so-called protracted ENSO of the early 1990s.

Madden–julian Oscillation

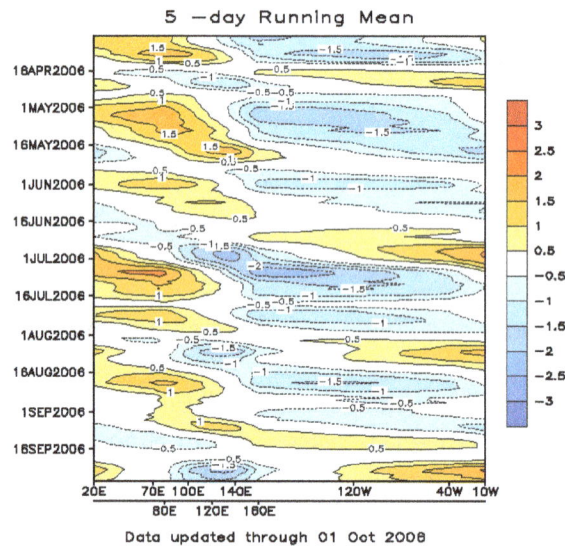

Note how the MJO moves eastward with time.

The Madden–Julian oscillation (MJO) is an equatorial traveling pattern of anomalous rainfall that is planetary in scale. It is characterized by an eastward progression of large regions of both enhanced and suppressed tropical rainfall, observed mainly over the Indian and Pacific Oceans. The anomalous rainfall is usually first evident over the western Indian Ocean, and remains evident as it propagates over the very warm ocean waters of the western and central tropical Pacific. This pattern of tropical rainfall then generally becomes very nondescript as it moves over the cooler ocean waters of the eastern Pacific but reappears over the tropical Atlantic and Indian Oceans. The wet phase of enhanced convection and precipitation is followed by a dry phase where convection is suppressed. Each cycle lasts approximately 30–60 days. The MJO is also known as the 30–60 day oscillation, 30–60 day wave, or the intraseasonal oscillation.

North Atlantic Oscillation (NAO)

Indices of the NAO are based on the difference of normalized sea level pressure (SLP) between Ponta Delgada, Azores and Stykkisholmur/Reykjavik, Iceland. The SLP anomalies at each station were normalized by division of each seasonal mean pressure by the long-term mean (1865–1984) standard deviation. Normalization is done to avoid the series of being dominated by the greater variability of the northern of the two stations. Positive values of the index indicate stronger-than-average westerlies over the middle latitudes.

Northern Annular Mode (NAM) or Arctic Oscillation (AO)

The NAM, or AO, is defined as the first EOF of northern hemisphere winter SLP data from the tropics and subtropics. It explains 23% of the average winter (December–March) variance, and it is dominated by the NAO structure in the Atlantic. Although there are some subtle differences from the regional pattern over the Atlantic and Arctic, the main difference is larger amplitude anomalies over the North Pacific of the same sign as those over the Atlantic. This feature gives the NAM a more annular (or zonally symmetric) structure.

Northern Pacific (NP) Index

The NP Index is the area-weighted sea level pressure over the region 30N–65N, 160E–140W.

Pacific Decadal Oscillation (PDO)

The PDO is a pattern of Pacific climate variability that shifts phases on at least inter-decadal time scale, usually about 20 to 30 years. The PDO is detected as warm or cool surface waters in the Pacific Ocean, north of 20° N. During a "warm", or "positive", phase, the west Pacific becomes cool and part of the eastern ocean warms; during a "cool" or "negative" phase, the opposite pattern occurs. The mechanism by which the pattern lasts over several years has not been identified; one suggestion is that a thin layer of warm water during summer may shield deeper cold waters. A PDO signal has been reconstructed to 1661 through tree-ring chronologies in the Baja California area.

Interdecadal Pacific Oscillation (IPO)

The Interdecadal Pacific oscillation (IPO or ID) display similar sea surface temperature (SST) and sea level pressure patterns to the PDO, with a cycle of 15–30 years, but affects both the north and south Pacific. In the tropical Pacific, maximum SST anomalies are found away from the equator. This is quite different from the quasi-decadal oscillation (QDO) with a period of 8–12 years and maximum SST anomalies straddling the equator, thus resembling ENSO.

Models

Climate models use quantitative methods to simulate the interactions of the atmosphere, oceans, land surface, and ice. They are used for a variety of purposes from study of the dynamics of the weather and climate system to projections of future climate. All climate models balance, or very nearly balance, incoming energy as short wave (including visible) electromagnetic radiation to the earth with outgoing energy as long wave (infrared) electromagnetic radiation from the earth. Any unbalance results in a change in the average temperature of the earth.

The most talked-about models of recent years have been those relating temperature to emissions of carbon dioxide. These models predict an upward trend in the surface temperature record, as well as a more rapid increase in temperature at higher latitudes.

Models can range from relatively simple to quite complex:

- A simple radiant heat transfer model that treats the earth as a single point and averages outgoing energy

- this can be expanded vertically (radiative-convective models), or horizontally

- finally, (coupled) atmosphere–ocean–sea ice global climate models discretise and solve the full equations for mass and energy transfer and radiant exchange.

Differences with Meteorology

In contrast to meteorology, which focuses on short term weather systems lasting up to a few weeks, climatology studies the frequency and trends of those systems. It studies the periodicity of weather events over years to millennia, as well as changes in long-term average weather patterns, in relation to atmospheric conditions. Climatologists study both the nature of climates – local, regional or global – and the natural or human-induced factors that cause climates to change. Climatology considers the past and can help predict future climate change.

Phenomena of climatological interest include the atmospheric boundary layer, circulation patterns, heat transfer (radiative, convective and latent), interactions between the atmosphere and the oceans and land surface (particularly vegetation, land use and topography), and the chemical and physical composition of the atmosphere.

Use in Weather Forecasting

A more complicated way of making a forecast, the analog technique requires remembering a previous weather event which is expected to be mimicked by an upcoming event. What makes it a difficult technique to use is that there is rarely a perfect analog for an event in the future. Some call this type of forecasting pattern recognition, which remains a useful method of observing rainfall over data voids such as oceans with knowledge of how satellite imagery relates to precipitation rates over land, as well as the forecasting of precipitation amounts and distribution in the future. A variation on this theme is used in Medium Range forecasting, which is known as teleconnections, when you use systems in other locations to help pin down the location of another system within the surrounding regime. One method of using teleconnections are by using climate indices such as ENSO-related phenomena.

References

- A. J. Bowden; Cynthia V. Burek; C. V. Burek; Richard Wilding (2005). History of palaeobotany: selected essays. Geological Society. p. 293. ISBN 978-1-86239-174-1. Retrieved 3 April 2013.

Evolution of Earth's Climate

The earth's climate is dependent on many factors such as volcanic eruptions and asteroid impacts, frequent during the early stages of the evolution of earth, to rising temperature levels caused by the emission of greenhouse gases. This chapter seeks to explain the recent additions to climate change records as well as instances of drastic climate change in the ancient past.

History of Climate Change Science

The history of the scientific discovery of climate change began in the early 19th century when ice ages and other natural changes in paleoclimate were first suspected and the natural greenhouse effect first identified. In the late 19th century, scientists first argued that human emissions of greenhouse gases could change the climate. Many other theories of climate change were advanced, involving forces from volcanism to solar variation. In the 1960s, the warming effect of carbon dioxide gas became increasingly convincing. Some scientists also pointed out that human activities that generated atmospheric aerosols (*e.g.,* "pollution") could have cooling effects as well. During the 1970s, scientific opinion increasingly favored the warming viewpoint. By the 1990s, as a result of improving fidelity of computer models and observational work confirming the Milankovitch theory of the ice ages, a consensus position formed: greenhouse gases were deeply involved in most climate changes and human caused emissions were bringing discernible global warming. Since the 1990s, scientific research on climate change has included multiple disciplines and has expanded. Research has expanded our understanding of causal relations, links with historic data and ability to model climate change numerically. Research during this period has been summarized in the Assessment Reports by the Intergovernmental Panel on Climate Change.

Climate change is a significant and lasting change in the statistical distribution of weather patterns over periods ranging from decades to millions of years. It may be a change in average weather conditions, or in the distribution of weather around the average conditions (such as more or fewer extreme weather events). Climate change is caused by factors that include oceanic processes (such as oceanic circulation), biotic processes, variations in solar radiation received by Earth, plate tectonics and volcanic eruptions, and human-induced alterations of the natural world. The latter effect is currently causing global warming, and "climate change" is often used to describe human-specific impacts.

Regional Changes, Antiquity Through 19th Century

From ancient times, people suspected that the climate of a region could change over the course of centuries. For example, Theophrastus, a pupil of Aristotle, told how the draining of marshes had made a particular locality more susceptible to freezing, and speculated that lands became warmer when the clearing of forests exposed them to sunlight. Renaissance and later scholars saw that deforestation, irrigation, and grazing had altered the lands around the Mediterranean since ancient times; they thought it plausible that these human interventions had affected the local weather. Vitruvius, in the first century BC, describes a series of ancient cities that lined the Anatolian peninsula from south to north along the Aegean sea, he then comments that they long ago were engulfed by the seas - presently these ancient cities are again out of water.

The most striking change came in the 18th and 19th centuries, obvious within a single lifetime: the conversion of Eastern North America from forest to croplands. By the early 19th century many believed the transformation was altering the region's climate—probably for the better. When sodbusters took over the Great Plains they were told that "rain follows the plough." Not everyone agreed. Some experts reported that deforestation not only caused rainwater to run off rapidly in useless floods, but reduced rainfall itself. European professors, alert to any proof that their nations were wiser than others, claimed that the Orientals of the Ancient Near East had heedlessly converted their once lush lands into impoverished deserts.

Meanwhile, national weather agencies had begun to compile masses of reliable observations of temperature, rainfall, and the like. When the figures were analyzed they showed many rises and dips, but no steady long-term change. By the end of the 19th century, scientific opinion had turned decisively against any belief in a human influence on climate. And whatever the regional effects, few imagined that humans could affect the climate of the planet as a whole.

Paleoclimate Change and Theories of its Causes, 19th Century

Erratics, boulders deposited by glaciers far from any existing glaciers, led geologists to the conclusion that climate had changed in the past.

Prior to the 18th century, scientists had not suspected that prehistoric climates were different from the modern period. By the late 18th century, geologists found evidence of a succession of geological ages with changes in climate. There were various competing theories about these changes, and James Hutton, whose ideas of cyclic change over huge periods of time were later dubbed uniformitarianism, was among those who found signs of past glacial activity in places too warm for glaciers in modern times.

In 1815 Jean-Pierre Perraudin described for the first time how glaciers might be responsible for the giant boulders seen in alpine valleys. As he hiked in the Val de Bagnes, he noticed giant granite rocks that were scattered around the narrow valley. He knew that it would take an exceptional force to move such large rocks. He also noticed how glaciers left stripes on the land, and concluded that it was the ice that had carried the boulders down into the valleys.

His idea was initially met with disbelief. Jean de Charpentier wrote, "I found his hypothesis so extraordinary and even so extravagant that I considered it as not worth examining or even considering." Despite Charpentier's initial rejection, Perraudin eventually convinced Ignaz Venetz that it might be worth studying. Venetz convinced Charpentier, who in turn convinced the influential scientist Louis Agassiz that the glacial theory had merit.

Agassiz developed a theory of what he termed "Ice Age" — when glaciers covered Europe and much of North America. In 1837 Agassiz was the first to scientifically propose that the Earth had been subject to a past ice age. William Buckland had led attempts in Britain to adapt the geological theory of catastrophism to account for erratic boulders and other "diluvium" as relics of the Biblical flood. This was strongly opposed by Charles Lyell's version of Hutton's uniformitarianism, and was gradually abandoned by Buckland and other catastrophist geologists. A field trip to the Alps with Agassiz in October 1838 convinced Buckland that features in Britain had been caused by glaciation, and both he and Lyell strongly supported the ice age theory which became widely accepted by the 1870s.

In the same general period that scientists first suspected climate change and ice ages, Joseph Fourier, in 1824, found that Earth's atmosphere kept the planet warmer than would be the case in a vacuum. Fourier recognized that the atmosphere transmitted visible light waves efficiently to the earth's surface. The earth then absorbed visible light and emitted infrared radiation in response, but the atmosphere did not transmit infrared efficiently, which therefore increased surface temperatures. He also suspected that human activities could influence climate, although he focused primarily on land use changes. In an 1827 paper Fourier stated, *"The establishment and progress of human societies, the action of natural forces, can notably change, and in vast regions, the state of the surface, the distribution of water and the great movements of the air. Such effects are able to make to vary, in the course of many centuries, the average degree of heat; because the analytic expressions contain coefficients relating to the state of the surface and which greatly influence the temperature."*

Eunice Newton Foote studied the warming effect of the sun, including how this warming was increased by the presence of carbonic acid gas (carbon dioxide), and suggested that the surface of an Earth whose atmosphere was rich in this gas would have a higher temperature. Her work was presented by Prof. Joseph Henry at the American Association for the Advancement of Science meeting in August 1856.

John Tyndall took Fourier's work one step further in 1864 when he investigated the absorption of infrared radiation in different gases. He found that water vapor, hydrocarbons like methane (CH_4), and carbon dioxide (CO_2) strongly block the radiation. Some scientists suggested that ice ages and other great climate changes were due to changes in the amount of gases emitted in volcanism. But that was only one of many possible causes. Another obvious possibility was solar variation. Shifts in ocean currents also might explain many climate changes. For changes over millions of years, the raising and lowering of mountain ranges would change patterns of both winds and ocean currents.

Or perhaps the climate of a continent had not changed at all, but it had grown warmer or cooler because of polar wander (the North Pole shifting to where the Equator had been or the like). There were dozens of theories.

For example, in the mid 19th century, James Croll published calculations of how the gravitational pulls of the Sun, Moon, and planets subtly affect the Earth's motion and orientation. The inclination of the Earth's axis and the shape of its orbit around the Sun oscillate gently in cycles lasting tens of thousands of years. During some periods the Northern Hemisphere would get slightly less sunlight during the winter than it would get during other centuries. Snow would accumulate, reflecting sunlight and leading to a self-sustaining ice age. Most scientists, however, found Croll's ideas—and every other theory of climate change—unconvincing.

First Calculations of Human-induced Climate Change, 1896

In 1896 Svante Arrhenius calculated the effect of a doubling atmospheric carbon dioxide to be an increase in surface temperatures of 5-6 degrees Celsius.

By the late 1890s, American scientist Samuel Pierpoint Langley had attempted to determine the surface temperature of the Moon by measuring infrared radiation leaving the Moon and reaching the Earth. The angle of the Moon in the sky when a scientist took a measurement determined how much CO_2 and water vapor the Moon's radiation had to pass through to reach the Earth's surface, resulting in weaker measurements when the Moon was low in the sky. This result was unsurprising given that scientists had known about infrared radiation absorption for decades.

A Swedish scientist, Svante Arrhenius, used Langley's observations of increased infrared absorption where Moon rays pass through the atmosphere at a low angle, encountering more carbon dioxide (CO_2), to estimate an atmospheric cooling effect from a future decrease of CO_2. He realized that the cooler atmosphere would hold less water vapor (another greenhouse gas) and calculated the additional cooling effect. He also realized the cooling would increase snow and ice cover at high latitudes, making the planet reflect more sunlight and thus further cool down, as James Croll had hypothesized. Overall Arrhenius calculated that cutting CO_2 in half would suffice to produce an ice age. He further calculated that a doubling of atmospheric CO_2 would give a total warming of 5-6 degrees Celsius.

Further, Arrhenius colleague Professor Arvid Högbom, who was quoted in length in Arrhenius 1896 study *On the Influence of Carbonic Acid in the Air upon the Temperature of the Earth* had been attempting to quantify natural sources of emissions of CO_2 for purposes of understanding the global carbon cycle. Högbom found that estimated carbon production from industrial sources in the 1890s (mainly coal burning) was comparable with the natural sources. Arrhenius saw that this human emission of carbon would eventually lead to warming. However, because of the relatively low rate of CO_2 production in 1896, Arrhenius thought the warming would take thousands of years, and he expected it would be beneficial to humanity.

Paleoclimates and Sunspots, Early 1900s to 1950s

Arrhenius's calculations were disputed and subsumed into a larger debate over whether atmospheric changes had caused the ice ages. Experimental attempts to measure infrared absorption in the laboratory seemed to show little differences resulted from increasing CO_2 levels, and also found significant overlap between absorption by CO_2 and absorption by water vapor, all of which suggested that increasing carbon dioxide emissions would have little climatic effect. These early experiments were later found to be insufficiently accurate, given the instrumentation of the time. Many scientists also thought that the oceans would quickly absorb any excess carbon dioxide.

Other theories of the causes of climate change fared no better. The principal advances were in observational paleoclimatology, as scientists in various fields of geology worked out methods to reveal ancient climates. Wilmot H. Bradley found that annual varves of clay laid down in lake beds showed climate cycles. An Arizona astronomer, Andrew Ellicott Douglass, saw strong indications of climate change in tree rings. Noting that the rings were thinner in dry years, he reported climate effects from solar variations, particularly in connection with the 17th-century dearth of sunspots (the Maunder Minimum) noticed previously by William Herschel and others. Other scientists, however, found good reason to doubt that tree rings could reveal anything beyond random regional variations. The value of tree rings for climate study was not solidly established until the 1960s.

Through the 1930s the most persistent advocate of a solar-climate connection was astrophysicist Charles Greeley Abbot. By the early 1920s, he had concluded that the solar "constant" was misnamed: his observations showed large variations, which he connected with sunspots passing across the face of the Sun. He and a few others pursued the topic into the 1960s, convinced that sunspot variations were a main cause of climate change. Other scientists were skeptical. Nevertheless, attempts to connect the solar cycle with climate cycles were popular in the 1920s and 1930s. Respected scientists announced correlations that they insisted were reliable enough to make predictions. Sooner or later, every prediction failed, and the subject fell into disrepute.

Meanwhile, the Serbian engineer Milutin Milankovitch, building on James Croll's theory, improved the tedious calculations of the varying distances and angles of the Sun's radiation as the Sun and Moon gradually perturbed the Earth's orbit. Some observations of varves (layers seen in the mud covering the bottom of lakes) matched the prediction of a Milankovitch cycle lasting about 21,000 years. However, most geologists dismissed the astronomical theory. For they could not fit

Milankovitch's timing to the accepted sequence, which had only four ice ages, all of them much longer than 21,000 years.

In 1938 a British engineer, Guy Stewart Callendar, attempted to revive Arrhenius's greenhouse-effect theory. Callendar presented evidence that both temperature and the CO_2 level in the atmosphere had been rising over the past half-century, and he argued that newer spectroscopic measurements showed that the gas was effective in absorbing infrared in the atmosphere. Nevertheless, most scientific opinion continued to dispute or ignore the theory.

Increasing Concern, 1950s - 1960s

Better spectrography in the 1950s showed that CO_2 and water vapor absorption lines did not overlap completely. Climatologists also realized that little water vapor was present in the upper atmosphere. Both developments showed that the CO_2 greenhouse effect would not be overwhelmed by water vapor.

In 1955 Hans Suess's carbon-14 isotope analysis showed that CO_2 released from fossil fuels was not immediately absorbed by the ocean. In 1957, better understanding of ocean chemistry led Roger Revelle to a realization that the ocean surface layer had limited ability to absorb carbon dioxide. Also predicting the rise in levels of CO_2. Later being proven by Charles David Keeling. By the late 1950s, more scientists were arguing that carbon dioxide emissions could be a problem, with some projecting in 1959 that CO_2 would rise 25% by the year 2000, with potentially "radical" effects on climate. In 1960 Charles David Keeling demonstrated that the level of CO_2 in the atmosphere was in fact rising. Concern mounted year by year along with the rise of the "Keeling Curve" of atmospheric CO_2.

Another clue to the nature of climate change came in the mid-1960s from analysis of deep-sea cores by Cesare Emiliani and analysis of ancient corals by Wallace Broecker and collaborators. Rather than four long ice ages, they found a large number of shorter ones in a regular sequence. It appeared that the timing of ice ages was set by the small orbital shifts of the Milankovitch cycles. While the matter remained controversial, some began to suggest that the climate system is sensitive to small changes and can readily be flipped from a stable state into a different one.

Scientists meanwhile began using computers to develop more sophisticated versions of Arrhenius's calculations. In 1967, taking advantage of the ability of digital computers to integrate absorption curves numerically, Syukuro Manabe and Richard Wetherald made the first detailed calculation of the greenhouse effect incorporating convection (the "Manabe-Wetherald one-dimensional radiative-convective model"). They found that, in the absence of unknown feedbacks such as changes in clouds, a doubling of carbon dioxide from the current level would result in approximately 2 °C increase in global temperature.

By the 1960s, aerosol pollution ("smog") had become a serious local problem in many cities, and some scientists began to consider whether the cooling effect of particulate pollution could affect global temperatures. Scientists were unsure whether the cooling effect of particulate pollution or warming effect of greenhouse gas emissions would predominate, but regardless, began to suspect that human emissions could be disruptive to climate in the 21st century if not sooner. In his 1968 book *The Population Bomb*, Paul R. Ehrlich wrote, "the greenhouse effect is being enhanced now by the greatly increased level of carbon dioxide... [this] is being countered by low-level clouds generated by contrails, dust, and other contaminants... At the moment we cannot predict what the overall climatic results will be of our using the atmosphere as a garbage dump."

In 1969, NATO was the first candidate to deal with climate change on an international level. It was planned then to establish a hub of research and initiatives of the organization in the civil area, dealing with environmental topics as Acid Rain and the Greenhouse effect. The suggestion of US President Richard Nixon was not very successful with the administration of German Chancellor Kurt Georg Kiesinger. But the topics and the preparation work done on the NATO proposal by the German authorities gained international momentum, as the government of Willy Brandt started to apply them on the civil sphere instead.

Scientists Increasingly Predict Warming, 1970s

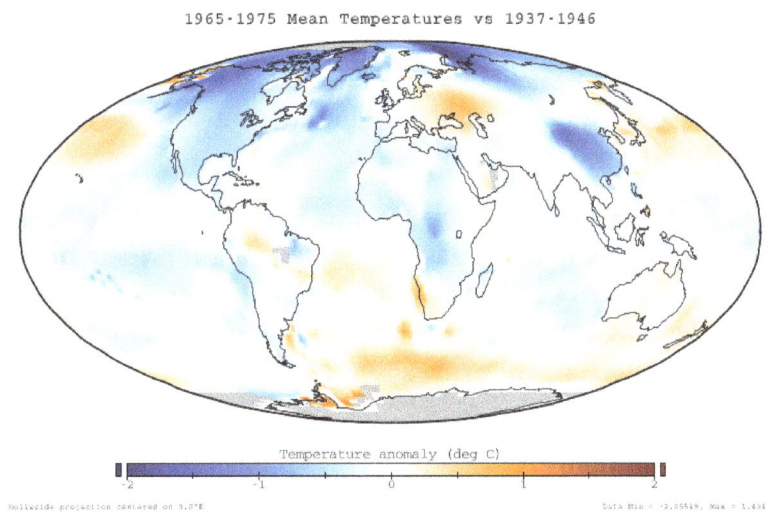

1965-1975 Mean Temperatures vs 1937-1946

Temperature anomaly (deg C)

Mean temperature anomalies during the period 1965 to 1975 with respect to the average temperatures from 1937 to 1946. This dataset was not available at the time.

In the early 1970s, evidence that aerosols were increasing world-wide encouraged Reid Bryson and some others to warn of the possibility of severe cooling. Meanwhile, the new evidence that the timing of ice ages was set by predictable orbital cycles suggested that the climate would gradually cool, over thousands of years. For the century ahead, however, a survey of the scientific literature from 1965 to 1979 found 7 articles predicting cooling and 44 predicting warming (many other articles on climate made no prediction); the warming articles were cited much more often in subsequent scientific literature. Several scientific panels from this time period concluded that more research was needed to determine whether warming or cooling was likely, indicating that the trend in the scientific literature had not yet become a consensus.

John Sawyer published the study *Man-made Carbon Dioxide and the "Greenhouse" Effect* in 1972. He summarized the knowledge of the science at the time, the anthropogenic attribution of the carbon dioxide greenhouse gas, distribution and exponential rise, findings which still hold today. Additionally he accurately predicted the rate of global warming for the period between 1972 and 2000.

The increase of 25% CO2 expected by the end of the century therefore corresponds to an increase of 0.6°C in the world temperature – an amount somewhat greater than the climatic variation of recent centuries. - John Sawyer, 1972

The mainstream news media at the time exaggerated the warnings of the minority who expected imminent cooling. For example, in 1975, *Newsweek* magazine published a story that warned of "ominous signs that the Earth's weather patterns have begun to change." The article continued by stating that evidence of global cooling was so strong that meteorologists were having "a hard time keeping up with it." On October 23, 2006, *Newsweek* issued an update stating that it had been "spectacularly wrong about the near-term future".

In the first two "Reports for the Club of Rome" in 1972 and 1974, the anthropogenic climate changes by CO_2 increase as well as by Waste heat were mentioned. About the latter John Holdren wrote in a study cited in the 1st report, *"... that global thermal pollution is hardly our most immediate environmental threat. It could prove to be the most inexorable, however, if we are fortunate enough to evade all the rest."* Simple global-scale estimates that recently have been actualized and confirmed by more refined model calculations show noticeable contributions from waste heat to global warming after the year 2100, if its growth rates are not strongly reduced (below the averaged 2% p.a. which occurred since 1973).

Evidence for warming accumulated. By 1975, Manabe and Wetherald had developed a three-dimensional Global climate model that gave a roughly accurate representation of the current climate. Doubling CO_2 in the model's atmosphere gave a roughly 2 °C rise in global temperature. Several other kinds of computer models gave similar results: it was impossible to make a model that gave something resembling the actual climate and not have the temperature rise when the CO_2 concentration was increased.

In a separate development, an analysis of deep-sea cores published in 1976 by Nicholas Shackleton and colleagues showed that the dominating influence on ice age timing came from a 100,000-year Milankovitch orbital change. This was unexpected, since the change in sunlight in that cycle was slight. The result emphasized that the climate system is driven by feedbacks, and thus is strongly susceptible to small changes in conditions.

In July 1979 the United States National Research Council published a report, concluding (in part):

When it is assumed that the CO_2 content of the atmosphere is doubled and statistical thermal equilibrium is achieved, the more realistic of the modeling efforts predict a global surface warming of between 2°C and 3.5°C, with greater increases at high latitudes.

... we have tried but have been unable to find any overlooked or underestimated physical effects that could reduce the currently estimated global warmings due to a doubling of atmospheric CO_2 to negligible proportions or reverse them altogether. ...

The 1979 World Climate Conference of the World Meteorological Organization concluded "it appears plausible that an increased amount of carbon dioxide in the atmosphere can contribute to a gradual warming of the lower atmosphere, especially at higher latitudes....It is possible that some effects on a regional and global scale may be detectable before the end of this century and become significant before the middle of the next century."

Consensus Begins to form, 1980-1988

By the early 1980s, the slight cooling trend from 1945-1975 had stopped. Aerosol pollution had decreased in many areas due to environmental legislation and changes in fuel use, and it became clear that the cooling effect from aerosols was not going to increase substantially while carbon dioxide levels were progressively increasing.

In 1982, Greenland ice cores drilled by Hans Oeschger, Willi Dansgaard, and collaborators revealed dramatic temperature oscillations in the space of a century in the distant past. The most prominent of the changes in their record corresponded to the violent Younger Dryas climate oscillation seen in shifts in types of pollen in lake beds all over Europe. Evidently drastic climate changes were possible within a human lifetime.

James Hansen during his 1988 testimony to Congress, which alerted the public to the dangers of global warming.

In 1973, British scientist James Lovelock speculated that chlorofluorocarbons (CFCs) could have a global warming effect. In 1975, V. Ramanathan found that a CFC molecule could be 10,000 times more effective in absorbing infrared radiation than a carbon dioxide molecule, making CFCs potentially important despite their very low concentrations in the atmosphere. While most early work on CFCs focused on their role in ozone depletion, by 1985 Ramanathan and others showed that CFCs together with methane and other trace gases could have nearly as important a climate effect as increases in CO_2. In other words, global warming would arrive twice as fast as had been expected.

In 1985 a joint UNEP/WMO/ICSU Conference on the "Assessment of the Role of Carbon Dioxide and Other Greenhouse Gases in Climate Variations and Associated Impacts" concluded that greenhouse gases "are expected" to cause significant warming in the next century and that some warming is inevitable.

Meanwhile, ice cores drilled by a Franco-Soviet team at the Vostok Station in Antarctica showed that CO_2 and temperature had gone up and down together in wide swings through past ice ages. This confirmed the CO_2-temperature relationship in a manner entirely independent of computer climate models, strongly reinforcing the emerging scientific consensus. The findings also pointed to powerful biological and geochemical feedbacks.

In June 1988, James E. Hansen made one of the first assessments that human-caused warming had already measurably affected global climate. Shortly after, a "World Conference on the Changing Atmosphere: Implications for Global Security" gathered hundreds of scientists and others in Toronto. They concluded that the changes in the atmosphere due to human pollution "represent a major threat to international security and are already having harmful consequences over many parts of the globe," and declared that by 2005 the world should push its emissions some 20% below the 1988 level.

The 1980s saw important breakthroughs with regard to global environmental challenges. E.g. Ozone depletion was mitigated by the Vienna Convention (1985) and the Montreal Protocol (1987). Acid rain was mainly regulated on the national and regional level.

Modern Period: 1988 to Present

In 1988 the WMO established the Intergovernmental Panel on Climate Change with the support of the UNEP. The IPCC continues its work through the present day, and issues a series of Assessment Reports and supplemental reports that describe the state of scientific understanding at the time each report is prepared. Scientific developments during this period are summarized about once every five to six years in the IPCC Assessment Reports which were published in 1990 (First Assessment Report), 1995 (Second Assessment Report), 2001 (Third Assessment Report), 2007 (Fourth Assessment Report), and 2013/2014 (Fifth Assessment Report).

Since the 1990s, research on climate change has expanded and grown, linking many fields such as atmospheric sciences, numeric modeling, behavioral sciences, geology and economics. Articles on climate change science are now frequently published in major journals such as Science by the American Association for the Advancement of Science and Nature, in addition there are focused journals on climate change research such as Nature Climate Change, Climate Change, Journal of Climate, Wiley Interdisciplinary Reviews: Climate Change, and International Journal of Climate Change Strategies and Management; furthermore, many journals on related subjects continue to publish articles that build out the science behind climate change (e.g. Quaternary Research).

Discovery of other Climate Changing Factors

Methane: In 1859, John Tyndall determined that coal gas, a mix of methane and other gases, strongly absorbed infrared radiation. Methane was subsequently detected in the atmosphere in 1948, and in the 1980s scientists realized that human emissions were having a substantial impact.

Chlorofluorocarbon: In 1973, British scientist James Lovelock speculated that chlorofluorocarbons (CFCs) could have a global warming effect. In 1975, V. Ramanathan found that a CFC molecule could be 10,000 times more effective in absorbing infrared radiation than a carbon dioxide molecule, making CFCs potentially important despite their very low concentrations in the atmosphere. While most early work on CFCs focused on their role in ozone depletion, by 1985 scientists had concluded that CFCs together with methane and other trace gases could have nearly as important a climate effect as increases in CO_2.

Timeline of Glaciation

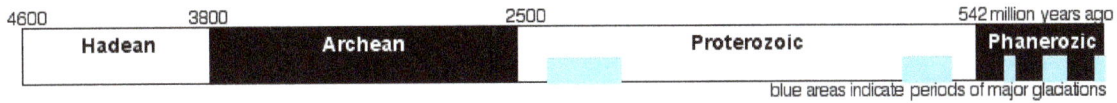

Showing the major glaciations (Ice Ages) in the context of Earth's entire existence.

There have been five known ice ages in the Earth's history, with the Earth experiencing the Quaternary Ice Age during the present time. Within ice ages, there exist periods of more severe glacial conditions and more temperate referred to as glacial periods and interglacial periods, respectively. The Earth is currently in such an interglacial period of the Quaternary Ice Age, with the last glacial period of the Quaternary having ended approximately 11,700 years ago with the start of the Holocene epoch. Based on climate proxies, paleoclimatologists study the different climate states originating from glaciation.

Known Ice Ages

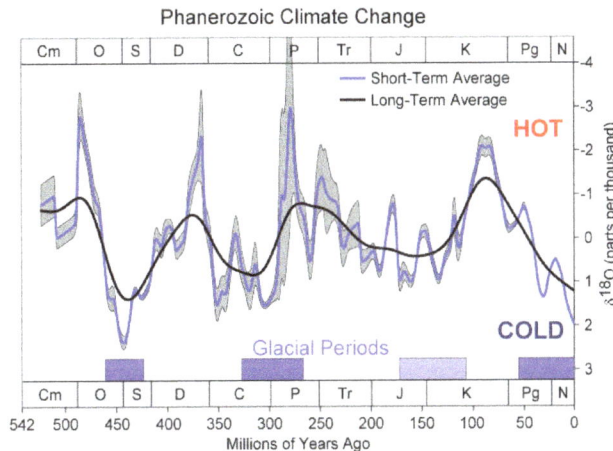

500 million year record shows current and previous two major glacial periods

Name	Period (Ma)	Period	Era
Quaternary	2.58 – present	Neogene	Cenozoic
Karoo	360 – 260	Carboniferous and Permian	Paleozoic
Andean-Saharan	450 – 420	Ordovician and Silurian	Paleozoic
Cryogenian (or Sturtian- Varangian)	850 – 635	Cryogenian	Neoproterozoic
Huronian	2400 – 2100	Siderian and Rhyacian	Paleoproterozoic

Descriptions

The second ice age, and possibly most severe, is estimated to have occurred from 850 to 635 Ma (million years) ago, in the Neoproterozoic Era and it has been suggested that it produced a second "Snowball Earth" in which the earth iced over completely. It has been suggested also that the end of this second cold period was responsible for the subsequent Cambrian Explosion, a time of rapid diversification of multicelled life during the Cambrian Period. However, this hypothesis is still controversial, though is growing in popularity among researchers as evidence in its favor has mounted.

A minor series of glaciations occurred from 460 Ma to 430 Ma. There were extensive glaciations from 350 to 250 Ma. The current ice age, called the Quaternary glaciation, has seen more or less extensive glaciation on 40,000 and later, 100,000 year cycles.

Nomenclature of Quaternary Glacial Cycles

Originally, the glacial and interglacial periods of the Quaternary Ice Age were named after characteristic geological features, and these names varied from region to region. It is now more common for researchers to refer to the periods by their marine isotopic stage number. The marine record preserves all the past glaciations; the land-based evidence is less complete because successive glaciations may wipe out evidence of their predecessors. Ice cores from continental ice accumulations also provide a complete record, but do not go as far back in time as marine data. Pollen data from lakes and bogs as well as loess profiles provided important land-based correlation data. The *names* system has not been completely filled out since the technical discussion moved to using marine isotopic stage numbers. For example, there are five Pleistocene glacial/interglacial cycles recorded in marine sediments during the last half million years, but only three classic interglacials were originally recognized on land during that period (Mindel, Riss and Würm).

Land-based evidence works acceptably well back as far as MIS 6, but it has been difficult to coordinate stages using just land-based evidence before that. Hence, the "names" system is incomplete and the land-based identifications of ice ages previous to that are somewhat conjectural. Nonetheless, land based data is essentially useful in discussing landforms, and correlating the known marine isotopic stage with them.

The last glacial and interglacial periods of the Quaternary are named, from most recent to most distant, as follows. Dates shown are in thousand years before present.

Land-based Chronology of Quaternary Glacial Cycles

Backwards Glacial Index	Names					Inter/Glacial	Period (ka)	Marine isotope stage (MIS)	Epoch
	Alpine	N. American	N. European	Great Britain	S. American				
				Flandrian		interglacial	present – 12	1	Holocene
1st	Würm	Wisconsin	Weichselian	Devensian	Llanquihue or Mérida	glacial period	12 – 71	2-4 & 5a-d	Pleistocene
	Riss-Würm	Sangamonian	Eemian	Ipswichian	Valdivia	interglacial	115 – 130	5e	
2nd	Riss	Illinoian	Saalian	Wolstonian or Gipping	Santa María	glacial period	130 – 200	6	
	Mindel-Riss	Pre-Illinoian	Holstein	Hoxnian		interglacial(s)	374 – 424	11	
3rd – 6th	Mindel	Pre-Illinoian	Elsterian	Anglian	Río Llico	glacial period(s)	424 – 478	12	
	Günz-Mindel	Pre-Illinoian		Cromerian*		interglacial(s)	478 – 533 – 563	13-15	
7th – 8th	Günz	Pre-Illinoian	Elbe or Menapian	Beestonian	Caracol	glacial period	621 – 676	16	

Older Periods of the Quaternary

Name	Inter/Glacial	Period (ka)	MIS	Epoch
Pastonian Stage	interglacial	600 – 800		
Pre-Pastonian Stage	glacial period	800 – 1300		
Bramertonian Stage	interglacial	1300 – 1550		

**Table data is based on Gibbard.

Ice Core Evidence of Recent Glaciation

Ice cores are used to obtain a high resolution record of recent glaciation. It confirms the chronology of the marine isotopic stages. Ice core data shows that the last 400,000 years have consisted of short interglacials (10,000 to 30,000 years) about as warm as the present alternated with much longer (70,000 to 90,000 years) glacials substantially colder than present. The new EPICA Antarctic ice core has revealed that between 400,000 and 780,000 years ago, interglacials occupied a considerably larger proportion of each glacial/interglacial cycle, but were not as warm as subsequent interglacials.

Piora Oscillation

The Piora Oscillation was an abrupt cold and wet period in the climate history of the Holocene Epoch; it is generally dated to the period of c. 3200 to 2900 BCE. Some researchers associate the Piora Oscillation with the end of the Atlantic climate regime, and the start of the Sub-Boreal, in the Blytt–Sernander sequence of Holocene climates.

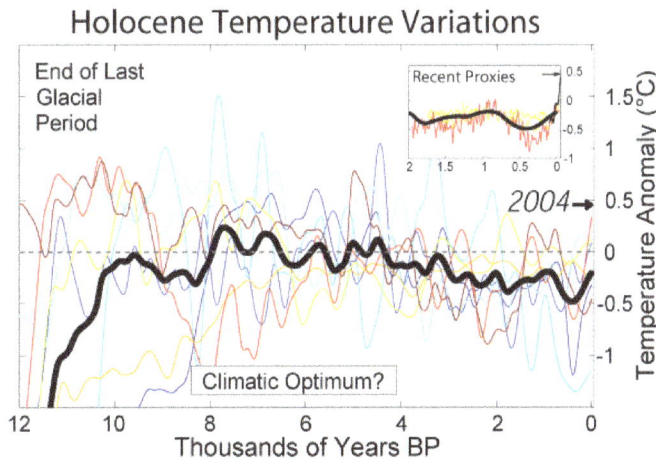

Holocene Temperature Variations

The spatial extent of the change is unclear; it does not show up as a major, or even identifiable, event in hemispheric temperature reconstructions.

First Detection

The phenomenon is named after the Val Piora or Piora Valley in Switzerland, where it was first detected; some of the most dramatic evidence of the Piora Oscillation comes from the region of the Alps. Glaciers advanced in the Alps, apparently for the first time since the Holocene climatic optimum; the Alpine tree line dropped by 100 meters. Yet the climate change extended far beyond the Alps and Europe. It affected what is now the New England region of North America, where hemlock and elm trees suffered a dramatic decline. Similar evidence comes from California and elsewhere; some changes in flora proved permanent. In the Middle East, the surface of the Dead Sea rose nearly 100 meters (300 feet), then receded to a more usual level. A few commentators have associated the climate changes of this period with the end of the Uruk period, as a Dark Age associated with the floods of the Gilgamesh epic and Noah's flood of the Book of Genesis.

Link with Horse Domestication

The Piora Oscillation has also been linked to the domestication of the horse. In Central Asia, a colder climate favored the use of horses: "The horse, since it was so adept at foraging with snow on the ground, tended to replace cattle and sheep." The Piora period seems associated with a period of colder drier air over the Western and Eastern Mediterranean, and may have depressed rainfalls as far afield as the Middle East. It is also associated with a sudden onset of drier weather in the central Sahara.

Causality

The cause or causes of the Piora Oscillation are debated. A Greenland ice core, GISP2, shows a sulfate spike and methane trough c. 3250 BCE, suggesting an unusual occurrence — either a volcanic eruption or a meteor or an asteroid impact event. Other authorities associate the Piora Oscillation with other comparable events, like the 8.2 kiloyear event, that recur in climate history, as part of a larger 1500-year climate cycle.

References

- Glacken, Clarence J. (1967). Traces on the Rhodian Shore. Nature and Culture in Western Thought from Ancient Times to the End of the Eighteenth Century. Berkeley: University of California Press. ISBN 978-0520032163.

- Fleming, James R. (1990). Meteorology in America, 1800-1870. Baltimore, MD: Johns Hopkins University Press. ISBN 978-0801839580.

- Young, Davis A. (1995). The biblical Flood: a case study of the Church's response to extrabiblical evidence. Grand Rapids, Mich: Eerdmans. ISBN 0-8028-0719-4. Retrieved 2008-09-16.

- David Archer (2009). The Long Thaw: How Humans Are Changing the Next 100,000 Years of Earth's Climate. Princeton University Press. p. 19. ISBN 978-0-691-13654-7.

- Lamb, Hubert H. (1997). Through All the Changing Scenes of Life: A Meteorologist's Tale. Norfolk, UK: Taverner. pp. 192–193. ISBN 1 901470 02 4.

- Fleming, James R. (2007). The Callendar Effect. The Life and Work of Guy Stewart Callendar (1898-1964), the Scientist Who Established the Carbon Dioxide Theory of Climate Change. Boston, MA: American Meteorological Society. ISBN 1878220764.

- Die Frühgeschichte der globalen Umweltkrise und die Formierung der deutschen Umweltpolitik(1950–1973) (Early history of the environmental crisis and the setup of German environmental policy 1950–1973), Kai F. Hünemörder, Franz Steiner Verlag, 2004 ISBN 3-515-08188-7

- H. Arnold, "Robert Döpel and his Model of Global Warming. An Early Warning – and its Update." Universitätsverlag Ilmenau (Germany) 2013. ISBN 978-3-86360 063-1.

- van Andel, Tjeerd H. (1994). New Views on an Old Planet: A History of Global Change (2nd ed.). Cambridge UK: Cambridge University Press. ISBN 0-521-44755-0.

- Gibbard, P.; van Kolfschoten, T. (2004). "Chapter 22: The Pleistocene and Holocene Epochs" (PDF). In Gradstein, F. M.; Ogg, James G.; Smith, A. Gilbert. A Geologic Time Scale 2004. Cambridge: Cambridge University Press. ISBN 0-521-78142-6.

- Matossian, Mary A. K. (1997). Shaping World History: Breakthroughs in Ecology, Technology, Science, and Politics. New York: M. E. Sharpe. ISBN 0-7656-0061-7.

- Sorenson, Raymond (January 11, 2011). "Eunice Foote's Pioneering Research On CO2 And Climate Warming" (PDF). Search and Discovery. AAPG/Datapages, Inc. (70092). Retrieved January 31, 2016.

- Foote, Eunice (November 1856). Circumstances affecting the Heat of the Sun's Rays. The American Journal of Science and Arts. pp. XXXI. Retrieved January 31, 2016.

- "Climate Science Milestones Leading To 1965 PCAST Report". Science. 350 (6264): 1046. November 27, 2015. doi:10.1126/science.350.6264.1046. Retrieved January 31, 2016.

- Brown, Dwayne; Cabbage, Michael; McCarthy, Leslie; Norton, Karen (20 January 2016). "NASA, NOAA Analyses Reveal Record-Shattering Global Warm Temperatures in 2015". NASA. Retrieved 21 January 2016.

3

An Overview of Climate

Climate is dependent on location and this demarcates the intensity with which a region experiences its climate. Climate defines the flora and fauna of a region, from tropical rainforests to desert regions. Climate is formed over a significant period of time. This chapter elucidates the crucial theories and principles of climatology.

Climate

Climate is the statistics (usually, mean or variability) of weather, usually over a 30-year interval. It is measured by assessing the patterns of variation in temperature, humidity, atmospheric pressure, wind, precipitation, atmospheric particle count and other meteorological variables in a given region over long periods of time. Climate differs from weather, in that weather only describes the short-term conditions of these variables in a given region.

A region's climate is generated by the climate system, which has five components: atmosphere, hydrosphere, cryosphere, lithosphere, and biosphere.

The climate of a location is affected by its latitude, terrain, and altitude, as well as nearby water bodies and their currents. Climates can be classified according to the average and the typical ranges of different variables, most commonly temperature and precipitation. The most commonly used classification scheme was Köppen climate classification originally developed by Wladimir Köppen. The Thornthwaite system, in use since 1948, incorporates evapotranspiration along with temperature and precipitation information and is used in studying biological diversity and the potential effects on it of climate changes. The Bergeron and Spatial Synoptic Classification systems focus on the origin of air masses that define the climate of a region.

Paleoclimatology is the study of ancient climates. Since direct observations of climate are not available before the 19th century, paleoclimates are inferred from *proxy variables* that include non-biotic evidence such as sediments found in lake beds and ice cores, and biotic evidence such as tree rings and coral. Climate models are mathematical models of past, present and future climates. Climate change may occur over long and short timescales from a variety of factors; recent warming is discussed in global warming.

Definition

Climate (from Ancient Greek *klima*, meaning *inclination*) is commonly defined as the weather averaged over a long period. The standard averaging period is 30 years, but other periods may be used depending on the purpose. Climate also includes statistics other than the average, such as the magnitudes of day-to-day or year-to-year variations. The Intergovernmental Panel on Climate Change (IPCC) 2001 glossary definition is as follows:

Climate in a narrow sense is usually defined as the "average weather," or more rigorously, as the statistical description in terms of the mean and variability of relevant quantities over a period ranging from months to thousands or millions of years. The classical period is 30 years, as defined by the World Meteorological Organization (WMO). These quantities are most often surface variables such as temperature, precipitation, and wind. Climate in a wider sense is the state, including a statistical description, of the climate system.

The World Meteorological Organization (WMO) describes climate "normals" as "reference points used by climatologists to compare current climatological trends to that of the past or what is considered 'normal'. A Normal is defined as the arithmetic average of a climate element (e.g. temperature) over a 30-year period. A 30 year period is used, as it is long enough to filter out any interannual variation or anomalies, but also short enough to be able to show longer climatic trends." The WMO originated from the International Meteorological Organization which set up a technical commission for climatology in 1929. At its 1934 Wiesbaden meeting the technical commission designated the thirty-year period from 1901 to 1930 as the reference time frame for climatological standard normals. In 1982 the WMO agreed to update climate normals, and in these were subsequently completed on the basis of climate data from 1 January 1961 to 31 December 1990.

The difference between climate and weather is usefully summarized by the popular phrase "Climate is what you expect, weather is what you get." Over historical time spans there are a number of nearly constant variables that determine climate, including latitude, altitude, proportion of land to water, and proximity to oceans and mountains. These change only over periods of millions of years due to processes such as plate tectonics. Other climate determinants are more dynamic: the thermohaline circulation of the ocean leads to a 5 °C (9 °F) warming of the northern Atlantic Ocean compared to other ocean basins. Other ocean currents redistribute heat between land and water on a more regional scale. The density and type of vegetation coverage affects solar heat absorption, water retention, and rainfall on a regional level. Alterations in the quantity of atmospheric greenhouse gases determines the amount of solar energy retained by the planet, leading to global warming or global cooling. The variables which determine climate are numerous and the interactions complex, but there is general agreement that the broad outlines are understood, at least insofar as the determinants of historical climate change are concerned.

Climate Classification

There are several ways to classify climates into similar regimes. Originally, climes were defined in Ancient Greece to describe the weather depending upon a location's latitude. Modern climate classification methods can be broadly divided into *genetic* methods, which focus on the causes of climate, and *empiric* methods, which focus on the effects of climate. Examples of genetic classification include methods based on the relative frequency of different air mass types or locations within synoptic weather disturbances. Examples of empiric classifications include climate zones defined by plant hardiness, evapotranspiration, or more generally the Köppen climate classification which was originally designed to identify the climates associated with certain biomes. A common shortcoming of these classification schemes is that they produce distinct boundaries between the zones they define, rather than the gradual transition of climate properties more common in nature.

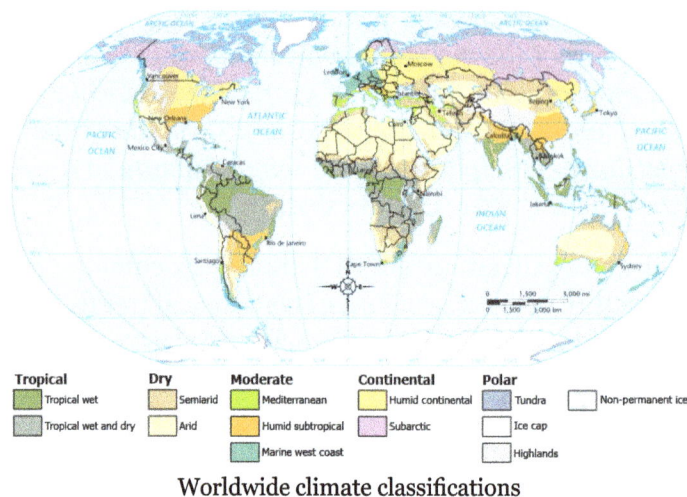

Worldwide climate classifications

Bergeron and Spatial Synoptic

The simplest classification is that involving air masses. The Bergeron classification is the most widely accepted form of air mass classification. Air mass classification involves three letters. The first letter describes its moisture properties, with c used for continental air masses (dry) and m for maritime air masses (moist). The second letter describes the thermal characteristic of its source region: T for tropical, P for polar, A for Arctic or Antarctic, M for monsoon, E for equatorial, and S for superior air (dry air formed by significant downward motion in the atmosphere). The third letter is used to designate the stability of the atmosphere. If the air mass is colder than the ground below it, it is labeled k. If the air mass is warmer than the ground below it, it is labeled w. While air mass identification was originally used in weather forecasting during the 1950s, climatologists began to establish synoptic climatologies based on this idea in 1973.

Based upon the Bergeron classification scheme is the Spatial Synoptic Classification system (SSC). There are six categories within the SSC scheme: Dry Polar (similar to continental polar), Dry Moderate (similar to maritime superior), Dry Tropical (similar to continental tropical), Moist Polar (similar to maritime polar), Moist Moderate (a hybrid between maritime polar and maritime tropical), and Moist Tropical (similar to maritime tropical, maritime monsoon, or maritime equatorial).

Köppen

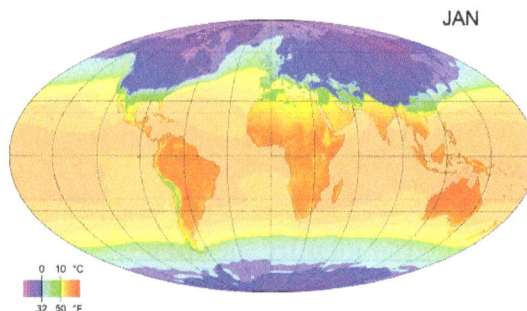

Monthly average surface temperatures from 1961–1990. This is an example of how climate varies with location and season

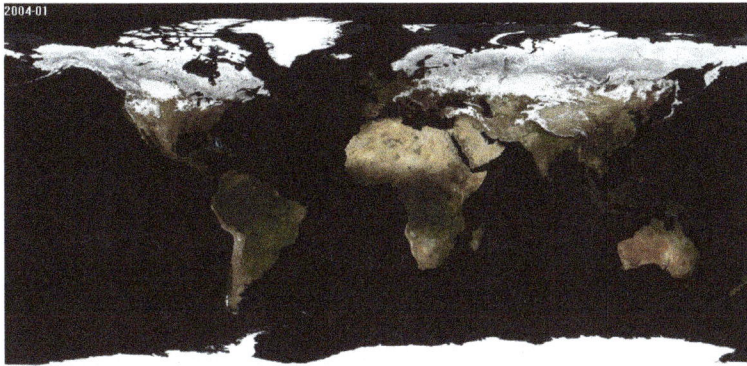

Monthly global images from NASA Earth Observatory (interactive SVG)

The Köppen classification depends on average monthly values of temperature and precipitation. The most commonly used form of the Köppen classification has five primary types labeled A through E. These primary types are A, tropical; B, dry; C, mild mid-latitude; D, cold mid-latitude; and E, polar. The five primary classifications can be further divided into secondary classifications such as rain forest, monsoon, tropical savanna, humid subtropical, humid continental, oceanic climate, Mediterranean climate, steppe, subarctic climate, tundra, polar ice cap, and desert.

Rain forests are characterized by high rainfall, with definitions setting minimum normal annual rainfall between 1,750 millimetres (69 in) and 2,000 millimetres (79 in). Mean monthly temperatures exceed 18 °C (64 °F) during all months of the year.

A monsoon is a seasonal prevailing wind which lasts for several months, ushering in a region's rainy season. Regions within North America, South America, Sub-Saharan Africa, Australia and East Asia are monsoon regimes.

The world's cloudy and sunny spots. NASA Earth Observatory map using data collected between July 2002 and April 2015.

A tropical savanna is a grassland biome located in semiarid to semi-humid climate regions of subtropical and tropical latitudes, with average temperatures remain at or above 18 °C (64 °F) year round and rainfall between 750 millimetres (30 in) and 1,270 millimetres (50 in) a year. They are widespread on Africa, and are found in India, the northern parts of South America, Malaysia, and Australia.

Cloud cover by month for 2014. NASA Earth Observatory

The humid subtropical climate zone where winter rainfall (and sometimes snowfall) is associated with large storms that the westerlies steer from west to east. Most summer rainfall occurs during thunderstorms and from occasional tropical cyclones. Humid subtropical climates lie on the east side continents, roughly between latitudes 20° and 40° degrees away from the equator.

Humid continental climate, worldwide

A humid continental climate is marked by variable weather patterns and a large seasonal temperature variance. Places with more than three months of average daily temperatures above 10 °C (50 °F) and a coldest month temperature below −3 °C (27 °F) and which do not meet the criteria for an arid or semiarid climate, are classified as continental.

An oceanic climate is typically found along the west coasts at the middle latitudes of all the world's continents, and in southeastern Australia, and is accompanied by plentiful precipitation year round.

The Mediterranean climate regime resembles the climate of the lands in the Mediterranean Basin, parts of western North America, parts of Western and South Australia, in southwestern South Africa and in parts of central Chile. The climate is characterized by hot, dry summers and cool, wet winters.

A steppe is a dry grassland with an annual temperature range in the summer of up to 40 °C (104 °F) and during the winter down to −40 °C (−40 °F).

A subarctic climate has little precipitation, and monthly temperatures which are above 10 °C (50 °F) for one to three months of the year, with permafrost in large parts of the area due to the cold winters. Winters within subarctic climates usually include up to six months of temperatures averaging below 0 °C (32 °F).

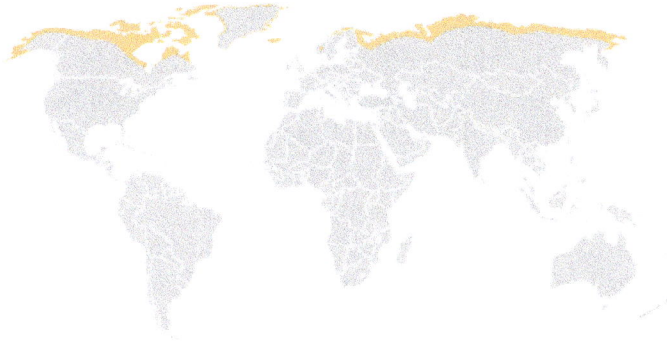

Map of arctic tundra

Tundra occurs in the far Northern Hemisphere, north of the taiga belt, including vast areas of northern Russia and Canada.

A polar ice cap, or polar ice sheet, is a high-latitude region of a planet or moon that is covered in ice. Ice caps form because high-latitude regions receive less energy as solar radiation from the sun than equatorial regions, resulting in lower surface temperatures.

A desert is a landscape form or region that receives very little precipitation. Deserts usually have a large diurnal and seasonal temperature range, with high or low, depending on location daytime temperatures (in summer up to 45 °C or 113 °F), and low nighttime temperatures (in winter down to 0 °C or 32 °F) due to extremely low humidity. Many deserts are formed by rain shadows, as mountains block the path of moisture and precipitation to the desert.

Thornthwaite

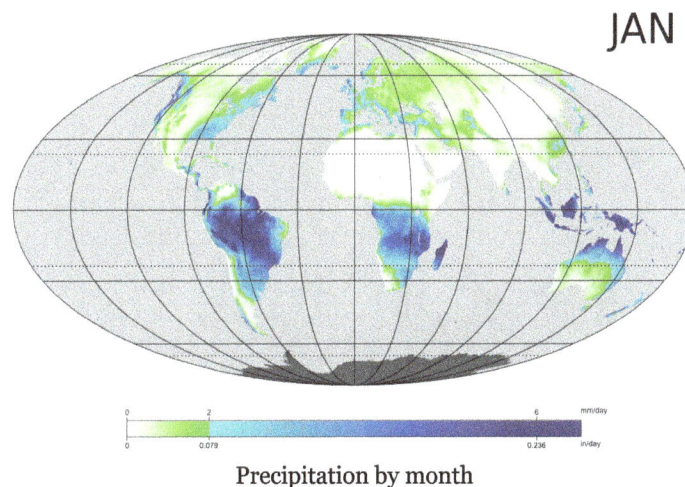

JAN

Precipitation by month

Devised by the American climatologist and geographer C. W. Thornthwaite, this climate classification method monitors the soil water budget using evapotranspiration. It monitors the portion of total precipitation used to nourish vegetation over a certain area. It uses indices such as a humidity index and an aridity index to determine an area's moisture regime based upon its average temperature, average rainfall, and average vegetation type. The lower the value of the index in any given area, the drier the area is.

The moisture classification includes climatic classes with descriptors such as hyperhumid, humid, subhumid, subarid, semi-arid (values of −20 to −40), and arid (values below −40). Humid regions experience more precipitation than evaporation each year, while arid regions experience greater evaporation than precipitation on an annual basis. A total of 33 percent of the Earth's landmass is considered either arid of semi-arid, including southwest North America, southwest South America, most of northern and a small part of southern Africa, southwest and portions of eastern Asia, as well as much of Australia. Studies suggest that precipitation effectiveness (PE) within the Thornthwaite moisture index is overestimated in the summer and underestimated in the winter. This index can be effectively used to determine the number of herbivore and mammal species numbers within a given area. The index is also used in studies of climate change.

Thermal classifications within the Thornthwaite scheme include microthermal, mesothermal, and megathermal regimes. A microthermal climate is one of low annual mean temperatures, generally between 0 °C (32 °F) and 14 °C (57 °F) which experiences short summers and has a potential evaporation between 14 centimetres (5.5 in) and 43 centimetres (17 in). A mesothermal climate lacks persistent heat or persistent cold, with potential evaporation between 57 centimetres (22 in) and 114 centimetres (45 in). A megathermal climate is one with persistent high temperatures and abundant rainfall, with potential annual evaporation in excess of 114 centimetres (45 in).

Record

Modern

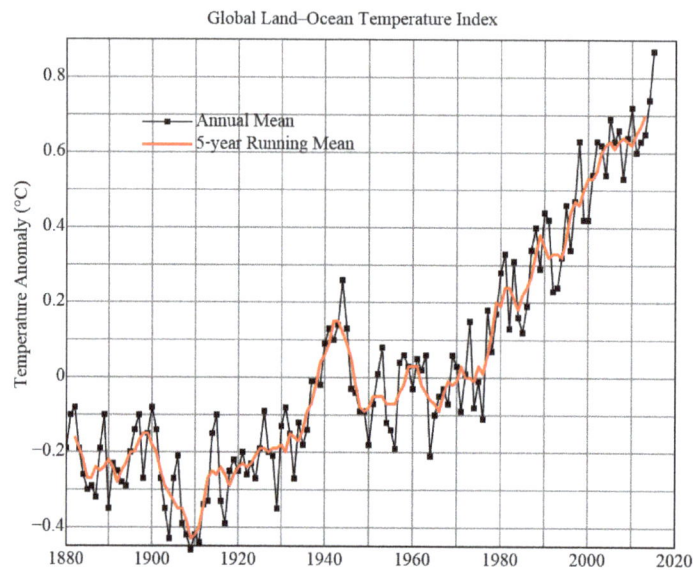

Global mean surface temperature change since 1880. Source: NASA GISS

Details of the modern climate record are known through the taking of measurements from such weather instruments as thermometers, barometers, and anemometers during the past few centuries. The instruments used to study weather over the modern time scale, their known error, their immediate environment, and their exposure have changed over the years, which must be considered when studying the climate of centuries past.

Paleoclimatology

Paleoclimatology is the study of past climate over a great period of the Earth's history. It uses evidence from ice sheets, tree rings, sediments, coral, and rocks to determine the past state of the climate. It demonstrates periods of stability and periods of change and can indicate whether changes follow patterns such as regular cycles.

Climate Change

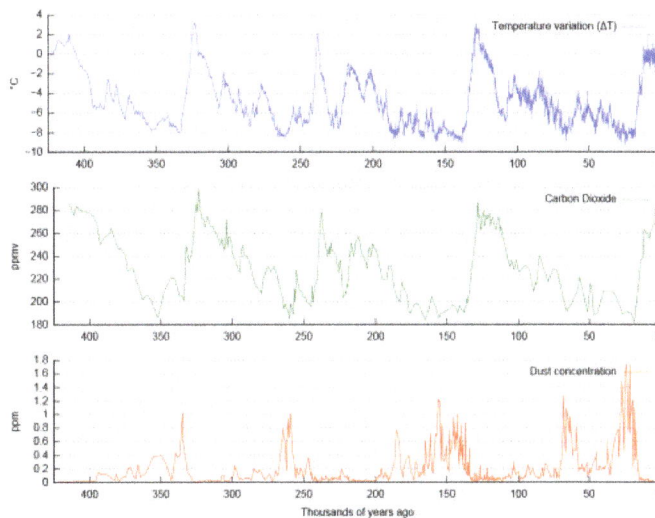

Variations in CO_2, temperature and dust from the Vostok ice core over the past 450,000 years

Climate change is the variation in global or regional climates over time. It reflects changes in the variability or average state of the atmosphere over time scales ranging from decades to millions of years. These changes can be caused by processes internal to the Earth, external forces (e.g. variations in sunlight intensity) or, more recently, human activities.

In recent usage, especially in the context of environmental policy, the term "climate change" often refers only to changes in modern climate, including the rise in average surface temperature known as global warming. In some cases, the term is also used with a presumption of human causation, as in the United Nations Framework Convention on Climate Change (UNFCCC). The UNFCCC uses "climate variability" for non-human caused variations.

Earth has undergone periodic climate shifts in the past, including four major ice ages. These consisting of glacial periods where conditions are colder than normal, separated by interglacial periods. The accumulation of snow and ice during a glacial period increases the surface albedo, reflecting more of the Sun's energy into space and maintaining a lower atmospheric temperature. Increases in greenhouse gases, such as by volcanic activity, can increase the global temperature and produce an interglacial period. Suggested causes of ice age periods include the positions of the continents, variations in the Earth's orbit, changes in the solar output, and volcanism.

Climate Models

Climate models use quantitative methods to simulate the interactions of the atmosphere, oceans, land surface and ice. They are used for a variety of purposes; from the study of the dynamics of the weather and climate system, to projections of future climate. All climate models balance, or very nearly balance, incoming energy as short wave (including visible) electromagnetic radiation to the earth with outgoing energy as long wave (infrared) electromagnetic radiation from the earth. Any imbalance results in a change in the average temperature of the earth.

The most talked-about applications of these models in recent years have been their use to infer the consequences of increasing greenhouse gases in the atmosphere, primarily carbon dioxide. These models predict an upward trend in the global mean surface temperature, with the most rapid increase in temperature being projected for the higher latitudes of the North-ern Hemisphere.

Models can range from relatively simple to quite complex:

- Simple radiant heat transfer model that treats the earth as a single point and averages out-going energy

- this can be expanded vertically (radiative-convective models), or horizontally

- finally, (coupled) atmosphere–ocean–sea ice global climate models discretise and solve the full equations for mass and energy transfer and radiant exchange.

Climate forecasting is a way by some scientists are using to predict climate change. In 1997 the prediction division of the International Research Institute for Climate and Society at Columbia University began generating seasonal climate forecasts on a real-time basis. To produce these forecasts an extensive suite of forecasting tools was developed, including a multimodel ensemble approach that required thorough validation of each model's accuracy level in simulating interannual climate variability.

Urban Climate

Urban climate refers to climatic conditions in an urban area that differ from neighboring rural areas, and are attributable to urban development. Urbanization tremendously changes the form of the landscape, and also produces changes in an area's air.

Temperatures

Temperatures are higher in cities than the surrounding rural areas—this area is called the urban heat island. There are a number of causes of the urban heat island:

- Building materials have a lower specific heat capacity (the amount of energy that will heat a kilogram of a material by 1 °C) than grass and trees—the specific heat capacity of concrete is 800 Joules/kg whereas for soil it can be 2000 Joules/kg, so concrete heats up more quickly in the day, warming the air around it.

- Buildings are heated, and vehicles and air conditioning systems generate heat.

- Buildings act as a barrier to winds which would otherwise distribute heat and cool the city.

- Buildings and roads with dark surfaces have a lower albedo (reflectivity) and absorb more sunlight, becoming hotter. Sunlight not absorbed by buildings is mostly reflected into other buildings.

The urban heat island effect tends to be stronger in winter because the colder air above the city is less able to rise by convection to allow the hot air inside the city to escape into the atmosphere. The effect is greater at night for the same reason.

Precipitation

Because cities are warmer, the hot air is more likely to rise and if it has a high humidity it will cause convectional rainfall – short intense bursts of rain and thunderstorms. Urban areas produce particles of dust (notably soot) and these act as hygroscopic nuclei which encourages rain production. Because of the warmer temperatures there is less snow in the city than surrounding areas.

Winds

Wind speeds are often lower in cities than the countryside because the buildings act as barriers (wind breaks). On the other hand, long streets with tall buildings can act as wind tunnels – winds funnelled down the street – and can be gusty as winds are channelled round buildings (eddying).

Humidity

Cities usually have a lower relative humidity than the surrounding air because cities are hotter, and rainwater in cities is unable to be absorbed into the ground to be released into the air by evaporation, and transpiration does not occur because cities have little vegetation. Surface runoff is usually taken up directly into the subterranean sewage water system and thus vanishes from the surface immediately.

Subnivean Climate

Subnivean climate (From Latin for "under" (sub) and "snow" (nives) refers to the zone in and underneath snowpack. This is the environment of many hibernal animals, as it provides protection from predators and insulation. The subnivean climate is formed by three different types of snow metamorphosis: destructive metamorphosis, which begins when snow falls; constructive metamorphosis, the movement of water vapor to the surface of the snowpack; and melt metamorphosis, the melting/sublimation of snow to water vapor and its refreezing in the snowpack. These three types of metamorphosis transform individual snowflakes into ice crystals and create spaces under the snow where small animals can move.

Subnivean Fauna

Subnivean fauna include small mammals such as mice, voles, shrews, and lemmings that must rely on winter snow cover for survival. These mammals move under the snow for protection from heat loss and some predators. In winter regions that do not have permafrost, the subnivean zone maintains a temperature of close to 32 °F (0 °C) regardless of the temperature above the snow cover, once the snow cover has reached a depth of six inches (15 cm) or more. The sinuous tunnels left by these small mammals can be seen from above when the snow melts to the final inch or so.

Some winter predators such as foxes and large owls can hear their prey through the snow and pounce from above. Ermine (stoats) can enter and hunt below the snowpack. Snowmobiles and ATVs can collapse the subnivean space. Skis and snow shoes are less likely to collapse subnivean space if the snowpack is deep enough.

Larger animals also utilize subnivean space. In the Arctic, ringed seals have closed spaces under the snow and above openings in the ice. In addition to resting and sleeping there, the female seals give birth to their pups on the ice. Female polar bears also den in snow caves to give birth to their young. Both types of dens are protected from exterior temperatures. Formation of these large spaces is from the animals' activity, not ground heat.

Subnivean Climate Formation

Deconstructive Metamorphosis

Deconstructive metamorphosis begins as the snow makes its way to the ground often melting, re-freezing, and settling. Water molecules become reordered causing the snowflakes to become more spherical in appearance. These melting snowflakes fuse with others around them becoming larger until all are uniform in size. While the snow is on the ground the melting and joining of snow flakes reduces the height of snowpack by shrinking air spaces and causing the density and mechanical strength of the snowpack to increase. Freshly fallen snow with a density of $0.1\,g/cm^3$ has very good insulating properties; however as time goes on, due to destructive metamorphism the insulating property of the snowpack decreases because the air spaces between snowflakes disappear. Snow that has been residing on the ground for a long period of time has an average density of $0.40\,g/cm^3$ and conducts heat well; however, once a base of 50 cm of snow with a density around $0.3\,g/cm^3$ has accumulated, temperatures under the snow remain relatively constant because the greater depth of snow compensates for its density. Destructive metamorphosis is a function of time, location, and weather. It occurs at a faster rate with higher temperatures, in the presence of water, under larger temperature gradients (e.g., warm days followed by cold nights), at lower elevations and on slopes that receive large amounts of solar radiation. As time goes on snow settles compacting air spaces, a process expedited by the packing force of the wind.

Compaction of snow reduces the penetration of long and short wave radiation by reflecting more radiation off the snow. This limitation of light transmission through the snowpack decreases light availability under the snow. Only three percent of light can penetrate to a depth of 20 cm of snow when the density is $0.21\,g/cm^3$. At a depth of 40 cm less than two tenths of a percent of light is transmitted from the snow surface to ground below. This decrease in light transmission occurs up to the point at which critical compaction is reached. This occurs because the surface area of the ice crystal decreases and it causes less refraction and scattering of light. Once densities reach 0.5 g/cm³, total surface area is reduced, which in turn reduces internal refraction and allows light to penetrate deeper into the snowpack.

Constructive Metamorphosis

Constructive metamorphosis is caused by the upward movement of water vapor within the snowpack. Warmer temperatures are found closer to the ground because it receives heat from the core of the earth. Snow has a low thermal conductivity so this heat is retained creating a temperature gradient between the air underneath the snowpack and the air above it. Warmer air holds more water vapor. Through the process of sublimation the newly formed water vapor travels vertically by way of diffusion from a higher concentration (next to the ground) to a lower concentration (near the snowpack surface) by traveling through the air spaces between ice crystals. When the water vapor reaches the top of the snowpack it is subjected to much colder air causing it to condense and refreeze, forming ice crystals at the top of the snowpack that can be seen as the layer of crust on top of the snow.

Melt Metamorphism

Melt metamorphism is the deterioration of snow by melting. Melting can be stimulated by warmer ambient temperatures, rain and fog. As snow melts water is formed and the force of gravity pulls these molecules downward. En route to the ground they refreeze thickening in the middle stratum. During this refreezing process energy is released in the form of latent heat. As more water comes down from the surface it creates more heat and brings the entire snowpack column to near equal temperature. The firnification of the snow strengthens the snowpack, due to the bonding of grains of snow. Snow around trees and under canopies melts faster due to the reradiation of long wave radiation. As snow gets older, particles of impurities (pine needles, dirt, and leaves, for example) accrue within the snow. These darkened objects absorb more short wave radiation causing them to rise in temperature, also reflecting more long wave radiation.

Köppen Climate Classification

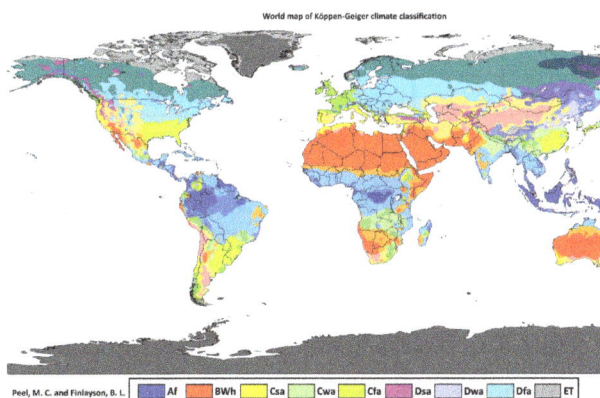

An updated Köppen–Geiger climate map

Köppen climate classification is one of the most widely used climate classification systems. It was first published by Russian German climatologist Wladimir Köppen in 1884, with several later modifications by Köppen, notably in 1918 and 1936. Later, German climatologist Rudolf Geiger (1954, 1961) collaborated with Köppen on changes to the classification system, which is thus sometimes called the Köppen–Geiger climate classification system.

The Köppen climate classification system has been further modified, within the Trewartha climate classification system in the middle 1960s (revised in 1980). The Trewartha system sought to create a more refined middle latitude climate zone, which was one of the criticisms of the Koppen system (the C climate group was too broad).

Scheme

Köppen climate classification scheme symbols description table.			
1st	**2nd**	**3rd**	**Description**
A			Tropical
	f		-Rainforest
	m		-Monsoon
	w		-Savanna
B			Arid
	W		-Desert
	S		-Steppe
		h	--Hot
		k	--Cold
C			Temperate
	s		-Dry Summer
	w		-Dry Winter
	f		-Without dry season
		a	--Hot Summer
		b	--Warm Summer
		c	--Cold Summer
D			Cold (Continental)
	s		-Dry Summer
	w		-Dry Winter
	f		-Without dry season
		a	--Hot Summer
		b	--Warm Summer
		c	--Cold Summer
		d	--Very cold Winter
E			Polar
	T		-Tundra
	F		-Frost (Ice cap)

The Köppen climate classification scheme divides climates into five main groups (A, B, C, D, E), each having several types and subtypes. Each particular climate type is represented by a two- to four-letter symbol.

Group A: Tropical/megathermal climates:

- Tropical rainforest climate (*Af*)

- Tropical monsoon climate (*Am*)

- Tropical wet and dry or savanna climate (*Aw*)

Group B: Dry (arid and semiarid) climates:

- Desert climate BW: Hot desert (*BWh*), Cold desert (*BWk*)

- Steppe climate (Semi-arid) BS: Hot steppe (*BSh*), Cold steppe (*BSk*)

Group C: Temperate/mesothermal climates:

- Dry-summer or Mediterranean climates (*Csa, Csb, Csc*)

- Temperate or subtropical hot-summer climates (*Cwa, Cfa*)

- Maritime temperate climates or Oceanic climates (*Cwb, Cwc, Cfb, Cfc*)

- Maritime subarctic climates or subpolar oceanic climate (*Cfc*)

- Temperate highland climates with dry winters (*Cwb, Cwc*)

Group D: Continental/microthermal climates

- Hot summer continental climates (*Dsa, Dwa, Dfa*)

- Warm summer continental or hemiboreal climates (*Dsb, Dwb, Dfb*)

- Continental subarctic or boreal (taiga) climates (*Dsc, Dwc, Dfc*)

- Continental subarctic climates with extremely severe winters (*Dsd, Dwd, Dfd*)

Group E: Polar and alpine climates:

- Tundra climate (*ET*)

- Ice cap climate (*EF*)

Meaning of symbols

These are the definitions from the 2007 version of the climate classification.

- T_{cold} – average (mean) temperature of coldest month

- T_{hot} – average (mean) temperature of hottest month

- T_{mon10} – number of months with average temperatures over 10 °C (50 °F)

- MAP – average (mean) annual precipitation

- MAT – average (mean) annual temperature

- summer half of year: whichever of October-March or April-September is hotter

- winter half of year: whichever of October-March or April-September is colder

- P_{swet} – average (mean) precipitation of wettest month in summer half of year

- P_{wwet} – average (mean) precipitation of wettest month in winter half of year

- P_{sdry} – average (mean) precipitation of driest month in summer half of year

- P_{wdry} – average (mean) precipitation of driest month in winter half of year

- $P_{threshold}$ –

    o if 70% of precipitation is in winter half of year, $2 \times MAT$

    o if 70% of precipitation is in summer half of year, $2 \times MAT + 28$

    o else $2 \times MAT + 14$

    •

    o A – $T_{cold} \geq 18\ °C\ (64\ °F)$

    o B – $MAP < 10 \times P_{threshold}$

        ☐ W – $MAP < 5 \times P_{threshold}$

        ☐ S – $MAP \geq 5 \times P_{threshold}$

            ☐ h – $MAT \geq 18\ °C\ (64\ °F)$

            ☐ k – $MAT < 18\ °C\ (64\ °F)$

    o C and D – $T_{hot} > 10\ °C\ (50\ °F)$

        ☐ C – $0\ °C\ (32\ °F) < T_{cold} < 18\ °C\ (64\ °F)$

        ☐ D – $T_{cold} \leq 0\ °C\ (32\ °F)$

            ☐ s – $P_{sdry} < 40\ mm\ (1.6\ in)$ and $P_{sdry} < P_{wwet}\ /\ 3$

            ☐ w – $P_{wdry} < P_{swet}\ /\ 10$

            ☐ f – not s or w

                ☐ a – $T_{hot} \geq 22\ °C\ (72\ °F)$

                ☐ b – T_{hot} not a and $T_{mon10} \geq 4$

                ☐ c – T_{hot} not a, b, or d and $1 \leq T_{mon10} < 4$

□ d (only D) – T_{cold} < –38 °C (–36 °F)

o E – T_{hot} < 10 °C (50 °F)

□ ET – T_{hot} > 0 °C (32 °F)

□ EF – T_{hot} < 0 °C (32 °F)

Group A: Tropical/megathermal climates

Tropical climates are characterized by constant high temperatures (at sea level and low elevations); all 12 months of the year have average temperatures of 18 °C (64.4 °F) or higher. They are subdivided as follows:

Tropical Rainforest Climate

All 12 months have average precipitation of at least 60 mm (2.4 in). These climates usually occur within 10° latitude of the equator. This climate is dominated by the doldrums low-pressure system all year round, so has no natural seasons. In some eastern-coast areas, they may extend to as much as 25° away from the equator when they share precipitation patterns with humid subtropical climates but feature warm enough temperatures to be classified as tropical.

- Examples:

  o Apia, Samoa

  o Bogor, Indonesia

  o Davao, Philippines

  o Innisfail, Queensland, Australia

  o Kisangani, Congo

  o Klang, Malaysia

  o Kuching, Malaysia

  o Paramaribo, Suriname

  o Quibdó, Colombia

  o Ratnapura, Sri Lanka

  o Singapore

  o West Palm Beach, Florida, United States

Some of the places with this climate are indeed uniformly and monotonously wet throughout the year (e.g., the northwest Pacific coast of South and Central America), but in many cases, the period of higher sun and longer days is distinctly wettest (as at Palembang, Indonesia) or the time of lower sun and shorter days may have more rain (as at Sitiawan, Malaysia).

A few places with this climate are found at the outer edge of the tropics, almost exclusively in the Southern Hemisphere; one example is Santos, São Paulo, Brazil.

(Note. The term aseasonal refers to the lack in the tropical zone of large differences in daylight hours and mean monthly (or daily) temperature throughout the year. Annual cyclic changes occur in the tropics, but not as predictably as those in the temperate zone, albeit unrelated to temperature, but to water availability whether as rain, mist, soil, or ground water. Plant response (e. g., phenology), animal (feeding, migration, reproduction, etc.), and human activities (plant sowing, harvesting, hunting, fishing, etc.) are tuned to this 'seasonality'. Indeed, in tropical South America and Central America, the 'rainy season' (and the 'high water season') is called *invierno* or *inverno*, though it could occur in the Northern Hemisphere summer; likewise, the 'dry season' (and 'low water season') is called *verano* or *verão*, and can occur in the Northern Hemisphere winter).

Tropical Monsoon Climate

This type of climate, most common in South America, results from the monsoon winds which change direction according to the seasons. This climate has a driest month (which nearly always occurs at or soon after the "winter" solstice for that side of the equator) with rainfall less than 60 mm, but more than 1/25 the total annual precipitation.

- Examples:

  - Abidjan, Ivory Coast

  - Cairns, Queensland, Australia

  - Chittagong, Bangladesh

  - Guanare, Portuguesa, Venezuela

  - Huế, Thừa Thiên–Huế, Vietnam

  - Jakarta, Indonesia

  - Macapá, Amapá, Brazil

  - Miami, Florida, United States

  - Port Harcourt, Rivers State, Nigeria

  - Puerto Ayacucho, Amazonas, Venezuela

  - Sinop, Mato Grosso, Brazil

  - Yangon, Myanmar

Also, another scenario exists under which some places fit into this category; this is referred to as the "trade-wind littoral" climate, because easterly winds bring enough precipitation during the "winter" months to prevent the climate from becoming a tropical wet-and-dry climate. Nassau, Bahamas, is included among these locations.

Tropical Wet and Dry or Savanna Climate

Aw climates have a pronounced dry season, with the driest month having precipitation less than 60 mm and less than 1/25 of the total annual precipitation.

- Examples:
    - Phnom Penh, Cambodia
    - Barquisimeto, Venezuela
    - Cartagena, Colombia
    - Ciudad Guayana, Venezuela
    - Dar es Salaam, Tanzania
    - Darwin, Northern Territory, Australia
    - Sanya, Hainan, China
    - Dili, East Timor
    - Ho Chi Minh City, Vietnam
    - Jijoca de Jericoacora, Ceará, Brazil
    - Kaohsiung, Taiwan
    - Kupang, Indonesia
    - Lagos, Lagos State, Nigeria
    - Medellín, Colombia
    - Mumbai, Maharashtra, India
    - Naples, Florida, United States
    - Port-au-Prince, Haiti
    - Rio de Janeiro, Rio de Janeiro, Brazil
    - Veracruz, Veracruz, Mexico

Most places that have this climate are found at the outer margins of the tropical zone from the low teens to the mid-20s latitudes, but occasionally an inner-tropical location (e.g., San Marcos, Antioquia, Colombia) also qualifies. Actually, the Caribbean coast, eastward from the Gulf of Urabá on the Colombia–Panamá border to the Orinoco River delta, on the Atlantic Ocean (about 4,000 km), have long dry periods (the extreme is the *BSh* climate, characterised by very low, unreliable precipitation, present, for instance, in extensive areas in the Guajira, and Coro, western Venezuela, the northernmost peninsulas in South America, which receive <300 mm total annual precipitation, practically all in two or three months).

This condition extends to the Lesser Antilles and Greater Antilles forming the circum-Caribbean dry belt. The length and severity of the dry season diminishes inland (southward); at the latitude of the Amazon River—which flows eastward, just south of the equatorial line—the climate is *Af*. East from the Andes, between the dry, arid Caribbean and the ever-wet Amazon are the Orinoco River's llanos or savannas, from where this climate takes its name.

Sometimes *As* is used in place of *Aw* if the dry season occurs during the time of higher sun and longer days. This is the case in parts of Hawaii, northwestern Dominican Republic (Monte Cristi, Villa Vásquez, Luperón), East Africa (Mombasa, Kenya), and Sri Lanka (Trincomalee), for instance. In most places that have tropical wet and dry climates, however, the dry season occurs during the time of lower sun and shorter days because of rain shadow effects during the 'high-sun' part of the year.

Group B: Dry (Arid and Semiarid) Climates

These climates are characterized by actual precipitation less than a threshold value set equal to the potential evapotranspiration. The threshold value (in millimeters) is determined as:

Multiply the average annual temperature in °C by 20, then add (a) 280 if 70% or more of the total precipitation is in the high-sun half of the year (April through September in the Northern Hemisphere, or October through March in the Southern), or (b) 140 if 30%–70% of the total precipitation is received during the applicable period, or (c) 0 if less than 30% of the total precipitation is so received.

According to the modified Köppen classification system used by modern climatologists, total precipitation in the warmest six months of the year is taken as reference instead of the total precipitation in the high-sun half of the year.

If the annual precipitation is less than 50% of this threshold, the classification is *BW* (arid: desert climate); if it is in the range of 50%–100% of the threshold, the classification is *BS* (semi-arid: steppe climate).

A third letter can be included to indicate temperature. Originally, *h* signified low-latitude climate (average annual temperature above 18 °C) while *k* signified middle-latitude climate (average annual temperature below 18 °C), but the more common practice today, especially in the United States, is to use *h* to mean the coldest month has an average temperature above 0 °C (32 °F), with *k* denoting that at least one month averages below 0 °C.

Desert areas situated along the west coasts of continents at tropical or near-tropical locations are characterized by cooler temperatures than encountered elsewhere at comparable latitudes (due to the nearby presence of cold ocean currents) and frequent fog and low clouds, despite the fact that these places rank among the driest on earth in terms of actual precipitation received. This climate is sometimes labelled *BWn*. The *BSn* category can be found in foggy coastal steppes.

- Arid climate examples:

    o Coober Pedy, Australia (*BWh*)

    o Baghdad, Iraq (*BWh*)

    o Upington, Northern Cape South Africa (*BWh*)

- o El Paso, Texas, United States (*BWh*)

- o Phoenix, Arizona, United States (*BWh*)

- o Death Valley, California, United States (*BWh*), location of the hottest air temperature ever recorded on Earth.

- o Las Vegas, Nevada, United States (*BWh*)

- o Hermosillo, Sonora, Mexico (*BWh*)

- o Almería, Andalusia, Spain (*BWh*)

- o 'Aziziya, Jafara, Libya (*BWh*)

- o Mecca, Makkah Region, Saudi Arabia (*BWh*)

- o Kuwait City, Capital Governorate, Kuwait (*BWh*)

- o Turpan, Xinjiang, China (*BWk*)

- o Nukus, Uzbekistan (*BWk*)

- o Lima, Peru (*BWn*)

- o Walvis Bay, Erongo Region, Namibia (*BWn*)

- Semi-arid examples:

  - o Lahore, Punjab, Pakistan (*BSh*)

  - o Aruba (*BSh*)

  - o Ivanhoe, New South Wales, Australia (*BSh*)

  - o Piraeus, Greece (*BSh*)

  - o Bamako, Mali (*BSh*)

  - o Ouagadougou, Burkina Faso (*BSh*)

  - o Niamey, Niger (*BSh*)

  - o N'Djamena, Chad (*BSh*)

  - o Yuanjiang Hani, Yi and Dai Autonomous County, Yunnan, China (*BSh*)

  - o Amman, Amman Governorate, Jordan (*BSk*)

  - o Baku, Azerbaijan (*BSk*)

  - o Yerevan, Armenia (*BSk*)

  - o Denver, Colorado, United States (*BSk*)

  - o Zacatecas City, Zacatecas, Mexico (*BSk*)

  - o Tabriz, East Azerbaijan Province, Iran (*BSk*)

o   Lethbridge, Alberta, Canada (*BSk*)

o   Brooks, Alberta, Canada (*BSk*)

o   Shijiazhuang, Hebei, China (*BSk*)

o   Lhasa, Tibet Autonomous Region, China (*BSk*)

o   Ulaanbaatar, Mongolia (*BSk*)

o   Comodoro Rivadavia, Chubut Province, Argentina (*BSk*)

On occasion, a fourth letter is added to indicate if either the winter or summer half of the year is "wetter" than the other half. To qualify, the wettest month must have at least 60 mm (2.4 in) of average precipitation if all 12 months are above 18 °C (64 °F), or 30 mm (1.2 in) if not; plus at least 70% of the total precipitation must be in the same half of the year as the wettest month — but the letter used indicates when the 'dry' season occurs, not the 'wet' one. This would result in Khartoum, Sudan, being reckoned as *BWhw*; Niamey, Niger, as *BShw*; Alexandria, Egypt, as *BWhs*; Asbi'ah, Libya, as *BShs*; Ömnögovi Province, Mongolia, as *BWkw*; and Xining, Qinghai, China, as *BSkw* (*BWks* and *BSks* do not exist if 0 °C in the coldest month is recognized as the *h/k* boundary.) If the standards for neither *w* nor *s* are met, no fourth letter is added.

Group C: Temperate/Mesothermal Climates

These climates have an average monthly temperature above 10 °C (50 °F) in their warmest months (April to September in northern hemisphere), and an average monthly temperature above −3 °C (27 °F) in their coldest months. Some climatologists prefer to observe 0 °C rather than −3 °C in the coldest month as the boundary between this group and the colder group D (continental).

In Asia, this includes areas from South Korea, to east- China from Beijing southward, to northern Japan. In Europe this includes areas from coastal Norway south to southern France, In the US, areas from near 40 latitude in the central and eastern states (a rough line from the NYC/NJ/CT area westward to the lower Ohio Valley, lower Midwest and southern Plains), are located in the Köppen C group.

The second letter indicates the precipitation pattern — *w* indicates dry winters (driest winter month average precipitation less than one-tenth wettest summer month average precipitation; one variation also requires that the driest winter month have less than 30 mm average precipitation), *s* indicates dry summers (driest summer month less than 40 mm average precipitation and less than one-third wettest winter month precipitation) and *f* means significant precipitation in all seasons (neither above-mentioned set of conditions fulfilled).

The third letter indicates the degree of summer heat — *a* indicates warmest month average temperature above 22 °C (72 °F) with at least four months averaging above 10 °C, *b* indicates warmest month averaging below 22 °C, but with at least four months averaging above 10 °C, while *c* means three or fewer months with mean temperatures above 10 °C.

The order of these two letters is sometimes reversed, especially by climatologists in the United States.

Mediterranean Climates

These climates usually occur on the western sides of continents between the latitudes of 30° and 45°. These climates are in the polar front region in winter, and thus have moderate temperatures and changeable, rainy weather. Summers are hot and dry, due to the domination of the subtropical high pressure systems, except in the immediate coastal areas, where summers are milder due to the nearby presence of cold ocean currents that may bring fog but prevent rain.

- Examples:
    - Beirut, Lebanon (*Csa*)
    - Latakia, Syria (*Csa*)
    - Tel Aviv, Israel (*Csa*)
    - Los Angeles, California, United States (*Csa* bordering a *Csb*)
    - Sacramento, California, United States (*Csa*)
    - Medford, Oregon, United States (*Csa*)
    - Perth, Australia (*Csa*)
    - Adelaide, Australia (*Csa*)
    - Tangier, Morocco(*Csa*)
    - Casablanca, Morocco (*Csa*)
    - Rome, Italy (*Csa*)
    - Seville, Spain (*Csa*)
    - Madrid, Spain (*Csa*)
    - Barcelona, Spain (*Csa*)
    - Marseille, France (*Csa*)
    - Nice, France (*Csa*)
    - Athens, Greece (*Csa*)
    - Antalya, Turkey (*Csa*)
    - Tashkent, Uzbekistan (*Csa*)
    - Dushanbe, Tajikistan (*Csa*)
    - Tunis, Tunisia (*Csa*)

- Dry-summer sometimes extends to additional areas not typically associated with a typical Mediterranean climate, as their warmest month mean doesn't reach 22 °C (71.6 °F), they are classified as (*Csb*). Some of these areas would border the Oceanic climate (*Cfb*), except their dry-summer patterns meet Köppen's *Cs* minimum thresholds.

- o   Examples include
    - ☐   Santiago, Chile (*Csb*)
    - ☐   Essaouira, Morocco (*Csb*)
    - ☐   Porto, Portugal (*Csb*)
    - ☐   San Francisco, California, United States (*Csb*)
    - ☐   Eureka, California, United States (*Csb*)
    - ☐   Portland, Oregon, United States (*Csb*)
    - ☐   Newport, Oregon, United States (*Csb*)
    - ☐   Seattle, Washington, United States (*Csb*)
    - ☐   Vancouver, British Columbia, Canada (*Csb*)
    - ☐   Cape Town, South Africa (*Csb*)
    - ☐   Kingston SE, Australia (*Csb*)
    - ☐   Albany, Western Australia, Australia (*Csb*)

- Cold-summer Mediterranean climates (*Csc*) exist in high-elevation areas adjacent to coastal *Csb* climate areas, where the strong maritime influence prevent the average winter monthly temperature from dropping below 0 °C. This climate is rare and is only found in climate fringes and isolated areas of the Cascades and Andes Mountains, as the dry-summer climate extends further poleward in the Americas than elsewhere.

    - o   Examples:
        - ☐   Balmaceda, Chile (*Csc*)
        - ☐   Cochamarca, Peru (*Csc*)
        - ☐   Røst, Norway (*Csc*)

Humid Subtropical Climates

These climates usually occur on the eastern coasts and eastern sides of continents, usually in the high 20s and 30s latitudes. Unlike the dry summer Mediterranean climates, humid subtropical climates have a warm and wet flow from the tropics that creates warm and moist conditions in the summer months. As such, summer (not winter as is the case in Mediterranean climates) is often the wettest season.

The flow out of the subtropical highs and the summer monsoon creates a southerly flow from the tropics that brings warm and moist air to the lower east sides of continents. This flow is often what brings the frequent but short-lived summer thundershowers so typical of the more southerly subtropical climates like the far southern United States, southern China and Japan.

- Examples:

- o   Buenos Aires, Argentina (*Cfa*)
- o   Porto Alegre, Brazil (*Cfa*)
- o   Brisbane, Australia (*Cfa*)
- o   Sydney, Australia (*Cfa*)
- o   Rasht, Gilan Province, Iran (*Cfa*)
- o   New Orleans, Louisiana, United States (*Cfa*)
- o   Orlando, Florida, United States (*Cfa*)
- o   Nashville, Tennessee, United States (*Cfa*)
- o   Washington D.C., United States (*Cfa*)
- o   Houston, Texas, United States (*Cfa*)
- o   Charleston, South Carolina, United States (*Cfa*)
- o   Atlanta, Georgia, United States (*Cfa*)
- o   Astara, Azerbaijan (*Cfa*)
- o   Horta, Azores (*Cfa*)
- o   Rize, Turkey (*Cfa*)
- o   Sochi, Russia (*Cfa*) – Warmest city to host a Winter Olympic Games.
- o   Kutaisi, Georgia (*Cfa*)
- o   Srinagar, India (*Cfa*)
- o   Shanghai, China (*Cfa*)
- o   Chengdu, Sichuan China (*Cfa*)
- o   Taipei, Taiwan (*Cfa*)
- o   Tokyo, Japan (*Cfa*)
- o   Milan, Italy (*Cfa*)
- o   Turin, Italy (*Cfa*)
- o   Tirana, Albania (*Cfa*)
- o   Toulouse, France (*Cfa*)

New York City is on the borderline between this climate and that of the humid continental (*Dfa*) climate, the line passing through the city.

- Subtropical-Dry Winter (*Cwa*) is monsoonal influenced, having the classic dry winter/wet summer pattern associated with tropical monsoonal climates.

- o Examples include:

    - ☐ Córdoba, Argentina (*Cwa*)

    - ☐ Santiago del Estero, Argentina (*Cwa*)

    - ☐ Mackay, Queensland, Australia (*Cwa*)

    - ☐ Islamabad, Pakistan (*Cwa*)

    - ☐ New Delhi, India (*Cwa*)

    - ☐ Haikou, Hainan China (*Cwa*)

    - ☐ Zhengzhou, Henan China (*Cwa*)

    - ☐ Xi'an, Shannxi China (*Cwa*)

    - ☐ Hong Kong (*Cwa*)

    - ☐ Kathmandu, Nepal (*Cwa*)

    - ☐ Imphal, India (*Cwa*)

    - ☐ Taunggyi, Myanmar (*Cwa*)

    - ☐ Hanoi, Vietnam (*Cwa*)

Oceanic Climates

Cfb climates usually occur in the higher middle latitudes on the western sides of continents between the latitudes of 40° and 60°; they are typically situated immediately poleward of the Mediterranean climates, although in Australia and extreme southern Africa this climate is found immediately poleward of temperate climates, and at a somewhat lower latitude. In western Europe, this climate occurs in coastal areas up to 63°N in Norway.

These climates are dominated all year round by the polar front, leading to changeable, often overcast weather. Summers are cool due to cool ocean currents, but winters are milder than other climates in similar latitudes, but usually very cloudy. *Cfb* climates are also encountered at high elevations in certain subtropical and tropical areas, where the climate would be that of a subtropical/tropical rain forest if not for the altitude. These climates are called "highlands".

- • Examples:

    - o Paris, France (*Cfb*)

    - o Bordeaux, France (*Cfb*)

    - o Bergen, Norway (*Cfb*)

    - o Berlin, Germany (*Cfb*)

    - o London, United Kingdom (*Cfb*)

- o Liverpool, England, United Kingdom (*Cfb*)

- o Southampton, England, United Kingdom (*Cfb*)

- o Shetland, Scotland (*Cfb*, bordering on *Cfc*)

- o Skagen, Denmark (*Cfb*, bordering on *Dfb* and *Cfc*)

- o Copenhagen, Denmark (*Cfb*, bordering on *Dfb*)

- o Zürich, Switzerland (*Cfb*)

- o Santander, Spain (*Cfb*)

- o Zonguldak, Turkey (*Cfb*)

- o Auckland, North Island, New Zealand (*Cfb*)

- o Mar del Plata, Buenos Aires Province, Argentina (*Cfb*)

- o Valdivia, Los Ríos Region, Chile (*Cfb*)

- o Quito, Pichincha Province, Ecuador (*Cfb*)

- o Bogotá, Colombia (*Cfb*)

- o Prince Rupert, British Columbia, Canada (*Cfb*)

- o Forks, Washington, United States (*Cfb*)

- o Ketchikan, Alaska, United States (*Cfb*)

- o Sky Valley, Georgia, United States (*Cfb*)

- o Block Island, Rhode Island, United States (*Cfb*)

- o Boone, North Carolina, United States (*Cfb*)

- o George, Western Cape, South Africa (*Cfb*)

- o Port Elizabeth, South Africa(Cfb)

- o Melbourne, Victoria, Australia (*Cfb*)

- o Hobart, Tasmania, Australia (*Cfb*)

- o Tristan da Cunha (*Cfb*)

- Subpolar oceanic climates (*Cfc*) occur poleward of or at higher elevations than the maritime temperate climates, and are mostly confined either to narrow coastal strips on the western poleward margins of the continents, or, especially in the Northern Hemisphere, to islands off such coasts. They occur most often from 60° to 70° latitude in both hemispheres.

  - o Examples:

- Grouse Mountain, North Vancouver, British Columbia

- Reykjavík, Iceland

- Unalaska, Alaska, United States

- Tórshavn, Faroe Islands

- Mount Read, Tasmania, Australia (borders on ET)

- Charlotte Pass, New South Wales, Australia

- Auckland Islands, New Zealand

Highland Climate

This is a type of climate mainly found in highlands inside the tropics of Mexico, Peru, Bolivia, Madagascar, Zambia, Zimbabwe and South Africa, but it is also found in central Argentina and areas outside of the tropics. Winters are noticeable and dry, and summers can be very rainy. In the tropics, the rainy season is provoked by the tropical air masses and the dry winters by subtropical high pressure.

- Examples:

    o Mexico City (*Cwb*)

    o Juliaca, Peru (*Cwb*)

    o Johannesburg, South Africa (*Cwb*)

    o Darjeeling, India (*Cwb*)

    o Kunming, Yunnan China (*Cwb*)

    o Weining, Guizhou China (*Cwb*)

    o Mokhotlong, Lesotho (*Cwb*)

    o Nairobi, Kenya (*Cwb*)

    o Copacabana, Bolivia (*Cwc*)

Group D: Continental/Microthermal Climates

These climates have an average temperature above 10 °C (50 °F) in their warmest months, and a coldest month average below −3 °C (or 0 °C in some versions, as noted previously). These usually occur in the interiors of continents and on their upper east coasts, normally north of 40°N. In the Southern Hemisphere, group D climates are extremely rare due to the smaller land masses in the middle latitudes and the almost complete absence of land at 40–60°S, existing only in some highland locations.

Group D climates are subdivided as:

The snowy city of Sapporo

Lettering

The second letter indicates the precipitation pattern — *w* indicates dry winters (driest winter month average precipitation less than one-tenth wettest summer month average precipitation; one variation also requires that the driest winter month have less than 30 mm average precipitation), *s* indicates dry summers (driest summer month less than 30 mm average precipitation and less than one-third wettest winter month precipitation) and *f* means significant precipitation in all seasons (neither above mentioned set of conditions fulfilled).

The third letter indicates the degree of summer heat and (for *c* and *d*) winter cold — *a* indicates warmest month average temperature above 22 °C (72 °F) with at least four months averaging above 10 °C, *b* indicates warmest month averaging below 22 °C, but with at least four months averaging above 10 °C, *c* indicates warmest month averaging below 22 °C and with three or fewer months with mean temperatures above 10 °C, but coldest month averaging above −38 °C, and *d* indicates warmest month averaging below 22 °C, three or fewer months averaging above 10 °C, and coldest month averaging below −38 °C.

Hot Summer Continental Climates

Dfa climates usually occur in the high 30s and low 40s latitudes, with a qualifying average temperature in the warmest month of >22 °C/72 °F. In Europe, these climates tend to be much drier than in North America. In eastern Asia, *Dwa* climates extend further south due to the influence of the Siberian high pressure system, which also causes winters there to be dry, and summers can be very wet because of monsoon circulation. *Dsa* exists at higher elevations adjacent to areas with hot summer Mediterranean (*Csa*) climates.

- Examples:

    o   Chicago, Illinois, United States (*Dfa*)

    o   Boston, Massachusetts, United States (*Dfa*)

    o   St. Louis, Missouri, United States (*Dfa*)

    o   Indianapolis, Indiana, United States (*Dfa*)

    o   Cleveland, Ohio, United States (*Dfa*)

    o   Sioux Falls, South Dakota, United States (*Dfa*)

    o   Omaha, Nebraska, United States (*Dfa*)

    o   Minneapolis, Minnesota, United States (*Dfa*, bordering on *Dfb*)

    o   Kansas City, Missouri, United States (*Dfa*)

    o   Pittsburgh, Pennsylvania, United States (*Dfa*)

    o   Cincinnati, Ohio, United States (*Dfa*)

    o   Windsor, Ontario, Canada/Detroit, Michigan, United States (*Dfa*)

    o   Bucharest, Romania (*Dfa*)

    o   Rostov-on-Don, Russia (*Dfa*)

    o   Almaty, Kazakhstan (*Dfa*)

    o   Aomori, Aomori Prefecture, Japan (*Dfa*)

    o   Nagano, Nagano Prefecture, Japan (*Dfa*)

    o   Pyongyang, North Korea (*Dwa*)

    o   Beijing, China (*Dwa*)

    o   Harbin, China (*Dwa*)

    o   Tianjin, China (*Dwa*)

    o   Changchun, China (*Dwa*)

    o   Seoul, South Korea (*Dwa*)

New York City is on the borderline between this climate and that of the humid subtropical (*Cfa*) climate, the line passing through the city.

- *Dsa* exists only at higher elevations adjacent to areas with hot summer Mediterranean (*Csa*) climates.

    o   Examples include:

        ▪   Sanandaj, Kurdistan Province, Iran

- Saqqez, Kurdistan Province, Iran
- Arak, Markazi Province, Iran
- Hakkâri, Turkey
- Muş, Turkey
- Bishkek, Kyrgyzstan
- Urzhar, Kazakhstan (bordering on *Dsb*)
- Salt Lake City, Utah, United States
- Cambridge, Idaho, United States
- Lytton, British Columbia, Canada

Warm summer continental or hemiboreal climates

Dfb and *Dwb* climates are immediately north of hot summer continental climates, generally in the high 40s and low 50s latitudes in North America and Asia, and also extending to higher latitudes in central and eastern Europe and Russia, between the maritime temperate and continental subarctic climates, where it extends up to 65 degrees latitude in places.

- Examples:
  - Helsinki, Finland (*Dfb*)
  - Svolvær, Lofoten, Nordland, Norway (*Dfb*, bordering on *Cfb*, *Cfc*, and *Dfc*)
  - Kiev, Ukraine (*Dfb*)
  - Stockholm, Sweden (*Dfb*)
  - Oslo, Norway (*Dfb*)
  - Portland, Maine, United States (*Dfb*)
  - Rochester, New York, United States (*Dfb*)
  - Binghamton, New York, United States (*Dfb*)
  - Lake Placid, New York, United States (*Dfb*)
  - Worcester, Massachusetts, United States (*Dfb*)
  - Youngstown, Ohio, United States (*Dfb*)
  - Calgary, Alberta, Canada (*Dfb*)
  - Saskatoon, Saskatchewan, Canada (*Dfb*)
  - Winnipeg, Manitoba, Canada (*Dfb*)
  - Ottawa, Ontario, Canada (*Dfb*)
  - London, Ontario, Canada (*Dfb*)

- o Edmonton, Alberta, Canada (*Dfb*)

- o Quebec City, Quebec, Canada (*Dfb*)

- o Montreal, Quebec, Canada (*Dfb*)

- o Moncton, New Brunswick, Canada (*Dfb*)

- o Halifax, Nova Scotia, Canada (*Dfb*)

- o Charlottetown, Prince Edward Island, Canada (*Dfb*)

- o St. John's, Newfoundland and Labrador, Canada (*Dfb*)

- o Erzurum, Turkey (*Dfb*)

- o Borjomi, Georgia (*Dfb*)

- o Warsaw, Poland (*Dfb*)

- o Prague, Czech Republic (*Dfb*)

- o Budapest, Hungary (*Dfb*)

- o Minsk, Belarus (*Dfb*)

- o Karaganda, Kazakhstan (*Dfb*)

- o Karaköl, Kyrgyzstan (*Dfb*)

- o Dickinson, North Dakota, United States (*Dwb*)

- o Heihe, China (*Dwb*)

- o Vladivostok, Russia (*Dwb*)

- o Khabarovsk, Russia (*Dwb*)

- o Pyeongchang County, South Korea (*Dwb*)

- *Dsb* arises from the same scenario as *Dsa*, but at even higher altitudes or latitudes, and chiefly in North America, since the Mediterranean climates extend further poleward than in Eurasia.

  - o Examples include:

    - Sivas, Turkey

    - Yüksekova, Turkey

    - Roghun, Tajikistan

    - Arslanbob, Kyrgyzstan

    - Flagstaff, Arizona, United States

    - South Lake Tahoe, California, United States

    - Bridgeport, California, United States

- Whistler, British Columbia, Canada

Subarctic or boreal climates

Dfc and *Dwc* climates occur poleward of the other group D climates, generally in the 50s and low 60s North latitudes. In some places, it extends northward to beyond 70°N latitude.

- Examples:
    - Murmansk, Murmansk Oblast, Russia (*Dfc*)
    - Yellowknife, Northwest Territories, Canada (*Dfc*)
    - Whitehorse, Yukon, Canada (*Dfc*)
    - Lillehammer, Norway (*Dfc*)
    - Churchill, Manitoba, Canada (*Dfc*)
    - Labrador City, Newfoundland and Labrador, Canada (*Dfc*)
    - Fraser, Colorado, United States (*Dfc*)
    - Fairbanks, Alaska, United States (*Dfc*)
    - Nome, Alaska, United States (*Dfc*)
    - Sütoluk, Ardahan Province, Turkey (*Dfc*)
    - St. Moritz, Grisons, Switzerland (*Dfc*)
    - Tignes, Savoie, France (*Dfc*)
    - Homer, Alaska, United States (*Dsc*)
    - Khandud, Badakhshan Province, Afghanistan (*Dsc*)
    - Bodie, California, United States (*Dsc*)
    - Bohemia Mountain, Oregon, United States (*Dsc*)
    - Spirit Lake, Washington, United States (*Dsc*), previous to 1980 eruption of Mount St. Helens
    - Mohe County, Heilongjiang, China (*Dwc*)
    - Yushu City, Qinghai, China (*Dwc*)
- Places with this climate (*Dfd* and *Dwd*) have severe winters, with the temperature in their coldest month lower than −38 °C. These climates occur only in eastern Siberia. The names of some of the places with this climate have become veritable synonyms for extreme, severe winter cold.
    - Examples:
        - Yakutsk, Sakha Republic, Russia (*Dfd*)

- Verkhoyansk, Sakha Republic, Russia (*Dfd*)

- Seymchan, Magadan Oblast, Russia (*Dfd*)

- Oymyakon, Sakha Republic, Russia (*Dwd*)

Group E: Polar and Alpine Climates

These climates are characterized by average temperatures below 10 °C in all 12 months of the year:

- Tundra climate (*ET*): Warmest month has an average temperature between 0 and 10 °C. These climates occur on the northern edges of the North American and Eurasian land masses, and on nearby islands. *ET* climates are also found on some islands near the Antarctic Convergence, and at high elevations outside the polar regions, above the tree line.

  o   Examples:

    - Macquarie Island (*ET*)

    - Crozet Islands (*ET*)

    - Campbell Island, New Zealand (*ET*)

    - Kerguelen Islands (*ET*)

    - Stanley, Falkland Islands (*ET*), borders subpolar oceanic (*Cfc*)

    - Ushuaia, Argentina (borders on *Cfc*)

    - La Rinconada, Peru (*ET*)

    - Mount Wellington, Australia (*ET*)

    - Murghab, Tajikistan (*ET*)

    - Nagqu, Tibet, China(*ET*)

    - Letseng diamond mine, Lesotho (*ET*, bordering on Cfc and Dfc)

  o   These climates are a colder and more continental variants of tundra. They would have characteristics of the ice cap climate, but still manage to see temperatures above 0 °C (32 °F):

    - Iqaluit, Nunavut, Canada (*ETf*)

    - Nanortalik, Greenland (*ETf*)

    - Mount Fuji, Japan (*ETf*)

    - Mount Washington, New Hampshire (*ETf*)

    - Eureka, Nunavut, Canada (*ETf*)

    - Nuuk, Greenland (*ETf*)

- Mys Shmidta, Russia (*ETf*)

- Dikson Island, Russia (*ETf*)

- Nord, Greenland, Greenland (*ETf*)

- Esperanza Base, Antarctica (*ETf*)

- Ice cap climate (*EF*): This climate is dominant in Antarctica and inner Greenland, but also occurs at extremely high altitudes on mountains, above even tundra. All twelve months have average temperatures below 0 °C (32 °F)

  o Examples:

    - Mount Rainier, Washington State (*EF*)

    - Mount Ararat, Turkey (*EF*)

    - Mount Everest, China/Nepal (*EF*)

    - Dye 3, Greenland (*EF*)

    - Summit Camp, Greenland (*EF*)

    - Scott Base, Antarctica (*EF*)

    - Vostok Station, Antarctica (*EF*), location of the lowest air temperature ever recorded on Earth.

    - McMurdo Station, Antarctica (*EF*)

    - Byrd Station, Antarctica (*EF*)

Occasionally, a third, lower-case letter is added to *ET* climates (distinguishing between *ETf*, *ETs*, and *ETw*), if either the summer or winter is clearly drier than the other half of the year. When the option to include this letter is exercised, the same standards that are used for Groups C and D apply, with the additional requirement that the wettest month must have an average of at least 30 mm precipitation (Group E climates can be as dry or even drier than Group B climates based on actual precipitation received, but their rate of evaporation is much lower). Seasonal precipitation letters are almost never attached to *EF* climates, mainly due to the difficulty in distinguishing between falling and blowing snow, as snow is the sole source of moisture in these climates.

Ecological Significance

The Köppen climate classification is based on the empirical relationship between climate and vegetation. This classification provides an efficient way to describe climatic conditions defined by temperature and precipitation and their seasonality with a single metric. Because climatic conditions identified by the Köppen classification are ecologically relevant, it has been widely used to map geographic distribution of long term climate and associated ecosystem conditions.

Over the recent years, there has been an increasing interest in using the classification to identify changes in climate and potential changes in vegetation over time. The most important ecological significance of the Köppen climate classification is that it helps to predict the dominant vegetation type based on the climatic data and vice versa.

In 2015, a pair of Chinese scholars published analysis of climate classifications between 1950 and 2010, finding that more than 5% of all land area worldwide had moved from wetter and colder classifications to drier and hotter classifications.

Trewartha Climate Classification Scheme

The Trewartha climate classification is a climate classification system published by American geographer Glenn Thomas Trewartha in 1966, and updated in 1980. It is a modified version of the 1899 Köppen system, created to answer some of the deficiencies of the Köppen system. The Trewartha system attempts to redefine the middle latitudes to be closer to vegetation zoning and genetic climate systems. It was considered a more true or "real world" reflection of the global climate.

For example, under the standard Köppen system, western Washington and Oregon are classed into the same climate zone as southern California, even though the two regions have strikingly different weather and vegetation. Under the old Köppen system cool oceanic climates like that of London were classed in the same zone as hot subtropical cities like Savannah, GA or Brisbane, Australia. In the US, locations in the Midwest like Ohio and Iowa which have long, severe winter climates where plants are completely dormant, were classed into the same climate zone as Louisiana or northern Florida which have mild winters and a green winter landscape.

Other Maps

All maps use the ≥0 °C definition for temperate climates and the 18 °C annual mean temperature threshold to distinguish between hot and cold dry climates.

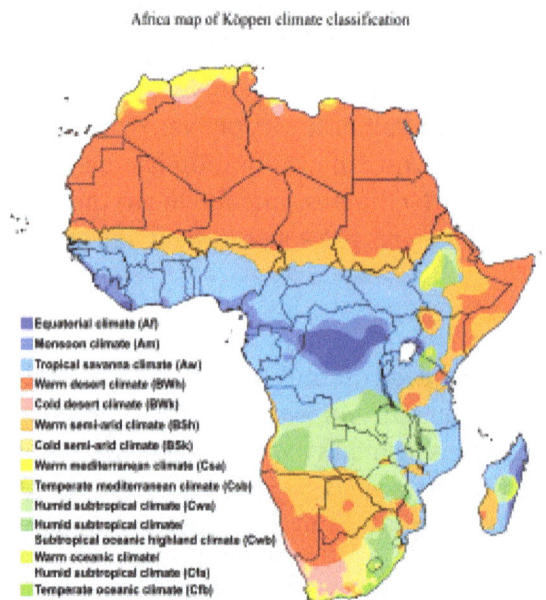

Köppen map of Africa

Central Asia map of Köppen climate classification

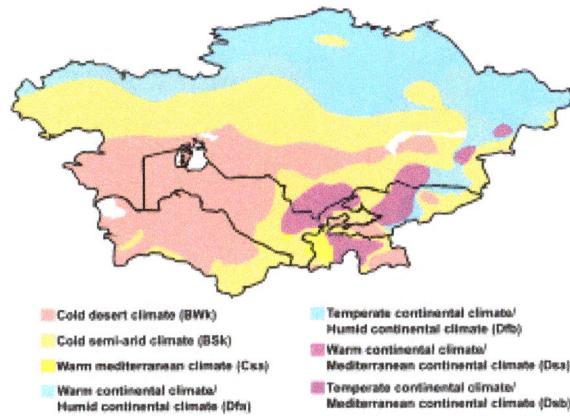

Köppen map of Central Asia

Oceania map of Köppen climate classification

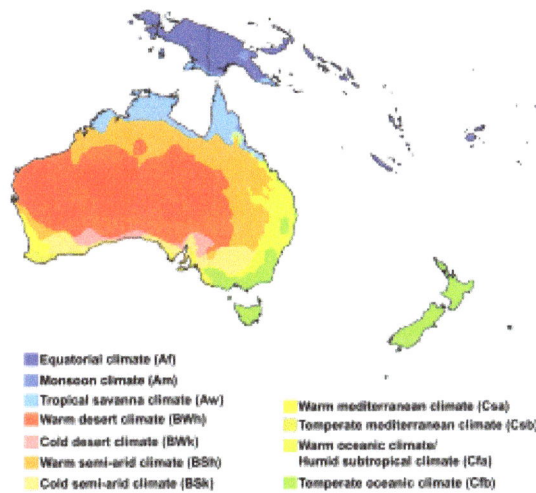

Köppen map of Australia/Oceania

South America map of Köppen climate classification

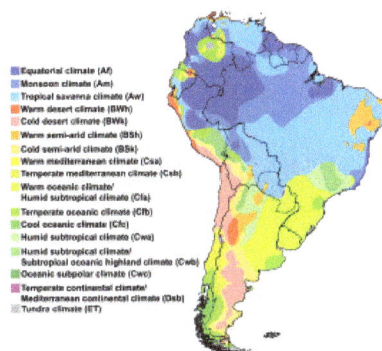

Köppen map of South America

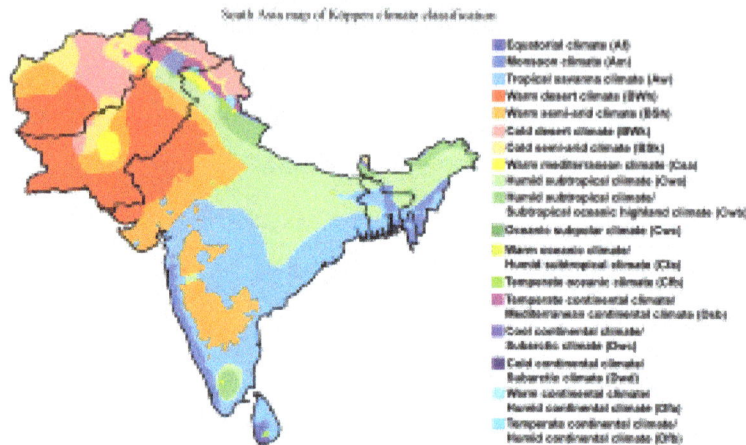

Köppen map of South Asia

References

- Fredlund, D.G.; Rahardjo, H. (1993). Soil Mechanics for Unsaturated Soils (PDF). Wiley-Interscience. ISBN 978-0-471-85008-3. OCLC 26543184. Retrieved 2008-05-21.

- McKnight, Tom L; Hess, Darrel (2000). "Climate Zones and Types". Physical Geography: A Landscape Appreciation. Upper Saddle River, NJ: Prentice Hall. ISBN 0-13-020263-0.

- Linacre, Edward; Bart Geerts (1997). Climates and Weather Explained. London: Routledge. p. 379. ISBN 0-415-12519-7.

- Planton, Serge (France; editor) (2013). "Annex III. Glossary: IPCC - Intergovernmental Panel on Climate Change" (PDF). IPCC Fifth Assessment Report. p. 1450. Retrieved 25 July 2016.

- Brown, Dwayne; Cabbage, Michael; McCarthy, Leslie; Norton, Karen (20 January 2016). "NASA, NOAA Analyses Reveal Record-Shattering Global Warm Temperatures in 2015". NASA. Retrieved 21 January 2016.

- Shepherd, Dr. J. Marshall; Shindell, Drew; O'Carroll, Cynthia M. (1 February 2005). "What's the Difference Between Weather and Climate?". NASA. Retrieved 13 November 2015.

- "Commission For Climatology: Over Eighty Years of Service" (pdf). World Meteorological Organization. 2011. pp. 6, 8, 10, 21, 26. Retrieved 1 September 2015.

- Gillis, Justin (28 November 2015). "Short Answers to Hard Questions About Climate Change". New York Times. Retrieved 29 November 2015.

Branches of Climatology

Climatology has gained popularity in recent years due to its ability to extract climate records from the past. Climatologists are able to compare as well as predict climate patterns, especially natural disasters and storms through the creation of climate models. This chapter is a compilation of the various branches of climatology that form an integral part of the broader subject matter.

Paleoclimatology

Paleoclimatology (in British spelling, palaeoclimatology) is the study of changes in climate taken on the scale of the entire history of Earth. It uses a variety of proxy methods from the Earth and life sciences to obtain data previously preserved within things such as rocks, sediments, ice sheets, tree rings, corals, shells and microfossils. It then uses the records to determine the past states of the Earth's various climate regions and its atmospheric system. Studies of past changes in the environment and biodiversity often reflect on the current situation, specifically the impact of climate on mass extinctions and biotic recovery.

History

The scientific study field of paleoclimate began to form in the early 19th century, when discoveries about glaciations and natural changes in Earth's past climate helped to understand the greenhouse effect.

Reconstructing Ancient Climates

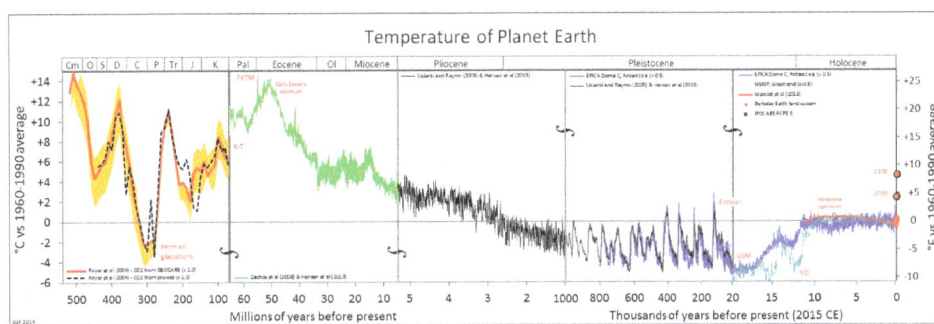

Palaeotemperature graphs compressed together

Paleoclimatologists employ a wide variety of techniques to deduce ancient climates.

Ice

Mountain glaciers and the polar ice caps/ice sheets provide much data in paleoclimatology.

Ice-coring projects in the ice caps of Greenland and Antarctica have yielded data going back several hundred thousand years, over 800,000 years in the case of the EPICA project.

- Air trapped within fallen snow becomes encased in tiny bubbles as the snow is compressed into ice in the glacier under the weight of later years' snow. The trapped air has proven a tremendously valuable source for direct measurement of the composition of air from the time the ice was formed.

- Layering can be observed because of seasonal pauses in ice accumulation and can be used to establish chronology, associating specific depths of the core with ranges of time.

- Changes in the layering thickness can be used to determine changes in precipitation or temperature.

- Oxygen-18 quantity changes ($\delta^{18}O$) in ice layers represent changes in average ocean surface temperature. Water molecules containing the heavier O-18 evaporate at a higher temperature than water molecules containing the normal Oxygen-16 isotope. The ratio of O-18 to O-16 will be higher as temperature increases. It also depends on other factors such as the water's salinity and the volume of water locked up in ice sheets. Various cycles in those isotope ratios have been detected.

- Pollen has been observed in the ice cores and can be used to understand which plants were present as the layer formed. Pollen is produced in abundance and its distribution is typically well understood. A pollen count for a specific layer can be produced by observing the total amount of pollen categorized by type (shape) in a controlled sample of that layer. Changes in plant frequency over time can be plotted through statistical analysis of pollen counts in the core. Knowing which plants were present leads to an understanding of precipitation and temperature, and types of fauna present. Palynology includes the study of pollen for these purposes.

- Volcanic ash is contained in some layers, and can be used to establish the time of the layer's formation. Each volcanic event distributed ash with a unique set of properties (shape and color of particles, chemical signature). Establishing the ash's source will establish a range of time to associate with layer of ice.

Dendroclimatology

Climatic information can be obtained through an understanding of changes in tree growth. Generally, trees respond to changes in climatic variables by speeding up or slowing down growth, which in turn is generally reflected a greater or lesser thickness in growth rings. Different species, however, respond to changes in climatic variables in different ways. A tree-ring record is established by compiling information from many living trees in a specific area.

Older intact wood that has escaped decay can extend the time covered by the record by matching the ring depth changes to contemporary specimens. By using that method, some areas have tree-ring records dating back a few thousand years. Older wood not connected to a contemporary record can be dated generally with radiocarbon techniques. A tree-ring record can be used to produce information regarding precipitation, temperature, hydrology, and fire corresponding to a particular area.

On a longer time scale, geologists must refer to the sedimentary record for data.

Sedimentary Content

- Sediments, sometimes lithified to form rock, may contain remnants of preserved vegetation, animals, plankton or pollen, which may be characteristic of certain climatic zones.

- Biomarker molecules such as the alkenones may yield information about their temperature of formation.

- Chemical signatures, particularly Mg/Ca ratio of calcite in Foraminifera tests, can be used to reconstruct past temperature.

- Isotopic ratios can provide further information. Specifically, the $\delta^{18}O$ record responds to changes in temperature and ice volume, and the $\delta^{13}C$ record reflects a range of factors, which are often difficult to disentangle.

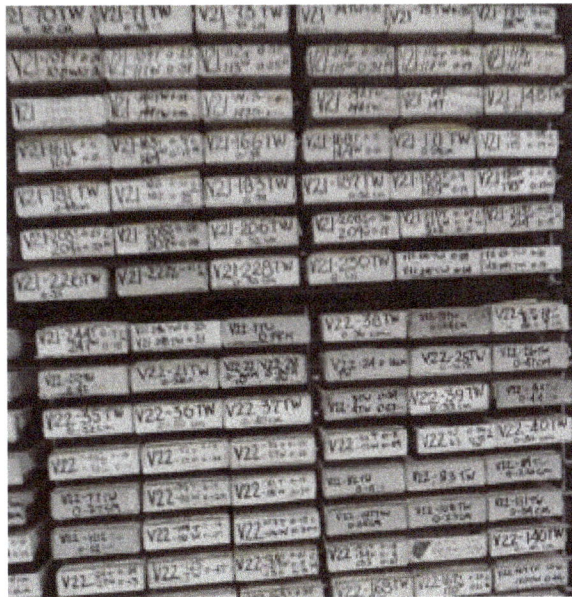

Sea floor core sample labelled to identify the exact spot on the sea floor where the sample was taken. Slight variations in location can make a significant difference in the chemical and biological composition of the sediment sample.

Sedimentary Facies

On a longer time scale, the rock record may show signs of sea level rise and fall, and features such as "fossilised" sand dunes can be identified. Scientists can get a grasp of long term climate by studying sedimentary rock going back billions of years. The division of earth history into separate periods is largely based on visible changes in sedimentary rock layers that demarcate major changes in conditions. Often, they include major shifts in climate.

Sclerochronology

Coral "rings" are similar to tree rings except that they respond to different things, such as the water temperature, freshwater influx, pH changes, and wave action. From there, certain equipment can be used to derive the sea surface temperature and water salinity from the past few centuries. The $\delta^{18}O$ of coralline red algae provides a useful proxy of the combined sea surface temperature and sea surface salinity at high latitudes and the tropics, where many traditional techniques are limited.

Limitations

A multinational consortium, the European Project for Ice Coring in Antarctica (EPICA), has drilled an ice core in Dome C on the East Antarctic ice sheet and retrieved ice from roughly 800,000 years ago. The international ice core community has, under the auspices of International Partnerships in Ice Core Sciences (IPICS), defined a priority project to obtain the oldest possible ice core record from Antarctica, an ice core record reaching back to or towards 1.5 million years ago. The deep marine record, the source of most isotopic data, exists only on oceanic plates, which are eventually subducted: the oldest remaining material is 200 million years old. Older sediments are also more prone to corruption by diagenesis. Resolution and confidence in the data decrease over time.

Notable Climate Events in Earth History

Knowledge of precise climatic events decreases as the record goes back in time, but some notable climate events are known

- Faint young Sun paradox (start)

- Huronian glaciation (~2400 Mya Earth completely covered in ice probably due to Great Oxygenation Event)

- Later Neoproterozoic Snowball Earth (~600 Mya, precursor to the Cambrian Explosion)

- Andean-Saharan glaciation (~450 Mya)

- Carboniferous Rainforest Collapse (~300 Mya)

- Permian–Triassic extinction event (251.4 Mya)

- Oceanic Anoxic Events (~120 Mya, 93 Mya, and others)

- Cretaceous–Paleogene extinction event (66 Mya)

- Paleocene–Eocene Thermal Maximum (Paleocene–Eocene, 55Mya)

- Younger Dryas/The Big Freeze (~11,000 BC)

- Holocene climatic optimum (~7000–3000 BC)

- Climate changes of 535-536 (535–536 AD)

- Medieval warm period (900–1300)

- Little Ice Age (1300–1800)

- Year Without a Summer (1816)

History of the Atmosphere

Earliest Atmosphere

The first atmosphere would have consisted of gases in the solar nebula, primarily hydrogen. In addition, there would probably have been simple hydrides such as those now found in gas giants like Jupiter and Saturn, notably water vapor, methane and ammonia. As the solar nebula dissipated, the gases would have escaped, partly driven off by the solar wind.

Second Atmosphere

The next atmosphere, consisting largely of nitrogen and carbon dioxide and inert gases, was produced by outgassing from volcanism, supplemented by gases produced during the late heavy bombardment of Earth by huge asteroids. A major part of carbon dioxide emissions were soon dissolved in water and built up carbonate sediments.

Water-related sediments have been found dating from as early as 3.8 billion years ago. About 3.4 billion years ago, nitrogen was the major part of the then stable "second atmosphere". An influence of life has to be taken into account rather soon in the history of the atmosphere because hints of early life forms have been dated to as early as 3.5 billion years ago. The fact that it is not perfectly in line with the 30% lower solar radiance (compared to today) of the early Sun has been described as the "faint young Sun paradox".

The geological record, however, shows a continually relatively warm surface during the complete early temperature record of Earth with the exception of one cold glacial phase about 2.4 billion years ago. In the late Archaean eon, an oxygen-containing atmosphere began to develop, apparent-ly from photosynthesizing cyanobacteria which have been found as stromatolite fossils from 2.7 billion years ago. The early basic carbon isotopy (isotope ratio propor-tions) was very much in line with what is found today, suggesting that the fundamental features of the carbon cycle were established as early as 4 billion years ago.

Third Atmosphere

The constant rearrangement of continents by plate tectonics influences the long-term evolution of the atmosphere by transferring carbon dioxide to and from large continental carbonate stores. Free oxygen did not exist in the atmosphere until about 2.4 billion years ago, during the Great Oxygenation Event, and its appearance is indicated by the end of the banded iron formations. Until then, any oxygen produced by photosynthesis was consumed by oxidation of reduced materials, notably iron. Molecules of free oxygen did not start to accumulate in the atmosphere until the rate of production of oxygen began to exceed the availability of reducing materials. That point was a shift from a reducing atmosphere to an oxidizing atmosphere. O_2 showed major variations until reaching a steady state of more than 15% by the end of the Precambrian. The following time span was the Phanerozoic eon, during which oxygen-breathing metazoan life forms began to appear.

The amount of oxygen in the atmosphere has fluctuated over the last 600 million years, reaching a peak of 35% during the Carboniferous period, significantly higher than today's 21%. Two main processes govern changes in the atmosphere: plants use carbon dioxide from the atmosphere, releasing oxygen and the breakdown of pyrite and volcanic eruptions release sulfur into the atmosphere, which oxidizes and hence reduces the amount of oxygen in the atmosphere. However, volcanic eruptions also release carbon dioxide, which plants can convert to oxygen. The exact cause of the variation of the amount of oxygen in the atmosphere is not known. Periods with much oxygen in the atmosphere are associated with rapid development of animals. Today's atmosphere contains 21% oxygen, which is high enough for rapid development of animals.

Currently, anthropogenic greenhouse gases are accumulating in the atmosphere, which is the main cause of global warming.

Climate During Geological Ages

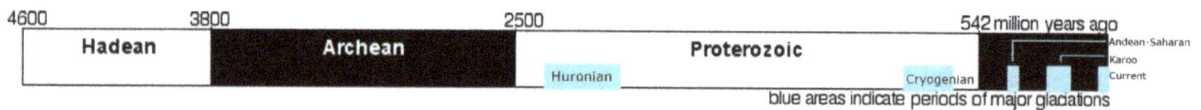

Timeline of glaciations, shown in blue

- The Huronian glaciation, is the first known glaciation in Earth's history, and lasted from 2400-2100 million years ago.

- The Cryogenian glaciation lasted from 850-635 million years ago.

- The Andean-Saharan glaciation lasted from 450–420 million years ago.

- The Karoo glaciation lasted from 360–260 million years ago.

- The Quaternary glaciation is the current glaciation period and begun 2.58 million years ago.

Precambrian Climate

The climate of the late Precambrian showed some major glaciation events spreading over much of the earth. At this time the continents were bunched up in the Rodinia supercontinent. Massive deposits of tillites are found and anomalous isotopic signatures are found, which gave rise to the Snowball Earth hypothesis. As the Proterozoic Eon drew to a close, the Earth started to warm up. By the dawn of the Cambrian and the Phanerozoic, life forms were abundant in the Cambrian explosion with average global temperatures of about 22 °C.

Phanerozoic Climate

Major drivers for the preindustrial ages have been variations of the sun, volcanic ashes and exhalations, relative movements of the earth towards the sun and tectonically induced effects as for major sea currents, watersheds and ocean oscillations. In the early Phanerozoic, increased atmospheric carbon dioxide concentrations have been linked to driving or amplifying increased global temperatures. Royer et al. 2004 found a climate sensitivity for the rest of the Phanerozoic which was calculated to be similar to today's modern range of values.

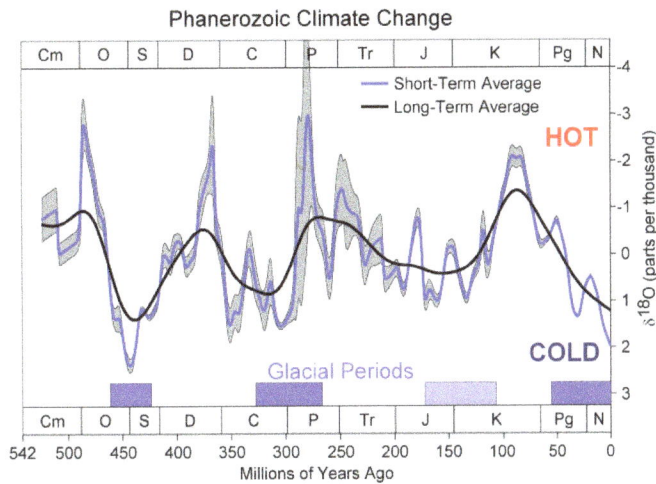

500 million years of climate change

The difference in global mean temperatures between a fully glacial Earth and an ice free Earth is estimated at approximately 10 °C, though far larger changes would be observed at high latitudes and smaller ones at low latitudes. One requirement for the development of large scale ice sheets seems to be the arrangement of continental land masses at or near the poles. The constant rearrangement of continents by plate tectonics can also shape long-term climate evolution. However, the presence or absence of land masses at the poles is not sufficient to guarantee glaciations or exclude polar ice caps. Evidence exists of past warm periods in Earth's climate when polar land masses similar to Antarctica were home to deciduous forests rather than ice sheets.

The relatively warm local minimum between Jurassic and Cretaceous goes along with an increase of subduction and mid-ocean ridge volcanism due to the breakup of the Pangea supercontinent .

Superimposed on the long-term evolution between hot and cold climates have been many short-term fluctuations in climate similar to, and sometimes more severe than, the varying glacial and interglacial states of the present ice age. Some of the most severe fluctuations, such as the Paleocene-Eocene Thermal Maximum, may be related to rapid climate changes due to sudden collapses of natural methane clathrate reservoirs in the oceans.

A similar, single event of induced severe climate change after a meteorite impact has been proposed as reason for the Cretaceous–Paleogene extinction event. Other major thresholds are the Permian-Triassic, and Ordovician-Silurian extinction events with various reasons suggested.

Quaternary Climate

The Quaternary sub-era includes the current climate. There has been a cycle of ice ages for the past 2.2–2.1 million years (starting before the Quaternary in the late Neogene Period).

Note in the graphic on the right the strong 120,000-year periodicity of the cycles, and the striking asymmetry of the curves. This asymmetry is believed to result from complex interactions of feedback mechanisms. It has been observed that ice ages deepen by progressive steps, but the recovery to interglacial conditions occurs in one big step.

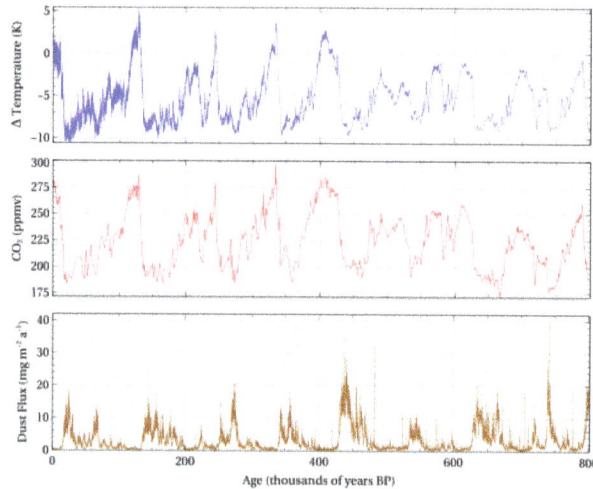

Ice core data for the past 800,000 years . Note length of glacial-interglacial cycles averages ~100,000 years. Blue curve is temperature, red curve is atmospheric CO_2 concentrations, and brown curve is dust fluxes. Today's date is on the left side of the graph because the x-axis values represent "age before 1950".

The graph below shows the temperature change over the past 12 000 years, from various sources. The thick black curve is an average.

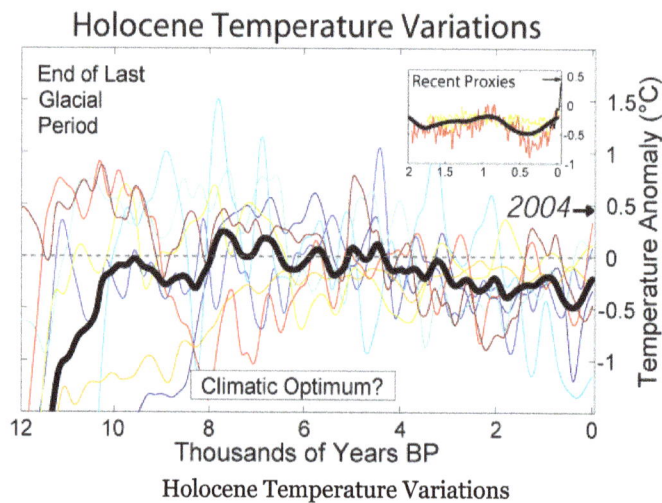

Holocene Temperature Variations

Climate Forcings

The climate forcing is the difference of radiant energy (sunlight) received by the Earth and the outgoing longwave radiation back to space. The radiative forcing is quantified based on the CO_2 amount in the tropopause, in units of watts per square meter to the Earth's surface. Dependent on the radiative balance of incoming and outgoing energy, the Earth either warms up or cools down. Earth radiative balance originates from changes in solar insolation and the concentrations of greenhouse gases and aerosols. Climate change may be due to internal processes in Earth sphere's and/or following external forcings.

Radiative forcing components

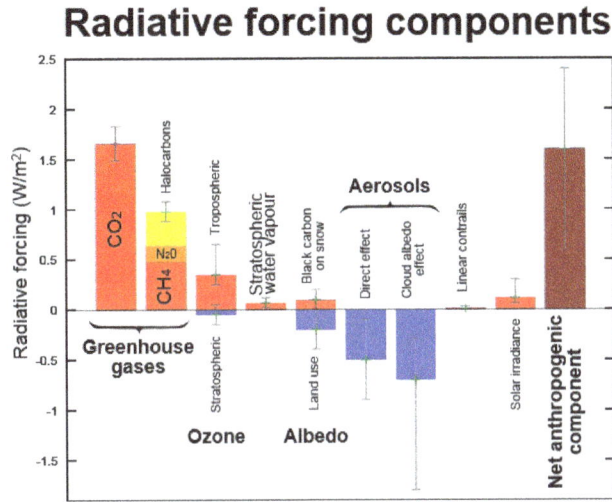

Radiative forcings, IPCC (2007)

Internal Processes and Forcings

The Earth's climate system involves the study of the atmosphere, biosphere, cryosphere, hydrosphere, and lithosphere, and the sum of these processes from Earth sphere's is considered the processes affecting the climate. Greenhouse gases act as the internal forcing of the climate system. Particular interests in climate science and paleoclimatology focuses on the study of Earth climate sensitivity, in response to the sum of forcings.

Examples:

- Thermohaline circulation (Hydrosphere)

- Life (Biosphere)

External Forcings

- The Milankovitch cycles determine Earth distance and position to the Sun. The solar insolation, is the total amount of solar radiation received by Earth.

- Volcanic eruptions, are considered an external forcing.

- Human changes of the composition of the atmosphere or land use.

Mechanisms

On timescales of millions of years, the uplift of mountain ranges and subsequently weathering processes of rocks and soils and the subduction of tectonic plates, are an important part of the carbon cycle. The weathering sequesters CO_2, by the reaction of minerals with chemicals (especially silicate weathering with CO_2) and thereby removing CO_2 from the atmosphere and reducing the radiative forcing. The opposite effect is volcanism, responsible for the natural greenhouse effect, by emitting CO_2 into the atmosphere, thus affecting glaciation (Ice Age) cycles.

James Hansen suggested that humans emit CO_2 10,000 times faster than natural processes have done in the past.

Ice sheet dynamics and continental positions (and linked vegetation changes) have been important factors in the long term evolution of the earth's climate. There is also a close correlation between CO_2 and temperature, where CO_2 has a strong control over global temperatures in Earth history.

Paleotempestology

Paleotempestology is the study of past tropical cyclone activity by means of geological proxies as well as historical documentary records. The term was coined by Kerry Emanuel.

Methods

Sedimentary Proxy Records

Examples of proxies include overwash deposits preserved in the sediments of coastal lakes and marshes, microfossils such as foraminifera, pollen, diatoms, dinoflagellates, phytoliths contained in coastal sediments, wave-generated or flood-generated sedimentary structures or deposits (called tempestites) in marine or lagoonal sediments, storm wave deposited coral shingle, shell, sand and shell and pure sand shore parallel ridges.

The method of using overwash deposits preserved in coastal lake and marsh sediments is adopted from earlier studies of paleotsunami deposits. Both storms and tsunamis leave very similar if not identical sedimentary deposits in coastal lakes and marshes and differentiating between the two in a sedimentary record can be difficult. The first studies to examine prehistoric records of tropical cyclones occurred in Australia and the South Pacific during the late 1970s and early 1980s. These studies examined multiple shore parallel ridges of coral shingle or sand and marine shells. As many as 50 ridges can be deposited at a site with each representing a past severe tropical cyclone over the previous 6,000 years. Tsunamis are not known to deposit multiple sedimentary ridges and therefore these features can be more easily attributed to a past storm at any given site.

Coastal sedimentary analyses have been done at the U.S. Gulf coast, the Atlantic coast from South Carolina up to New Jersey and New England, and the Caribbean Sea. Also, studies on pre-historic tropical cyclones hitting Australia have been made. A study covering the South China Sea coast has also been published.

Markers in Coral

Rocks contain certain isotopes of elements, known as natural tracers, which describe the conditions under which they formed. By studying the calcium carbonate in coral rock, past sea surface temperature and hurricane information can be revealed. Heavier oxygen isotopes (^{18}O) decrease in relation to lighter oxygen isotopes (^{16}O) in coral during periods of very heavy rainfall. Since hurricanes are the main source of extreme rainfall in the tropical oceans, past hurricane events can be dated to the days of their impact on the coral by looking at the decreased ^{18}O concentration within the coral.

Speleothems and Tree Rings

Isotope studies in speleothems and tree rings (dendrochronology) offers a means by which higher resolution records of long-term tropical cyclone histories can be attained. Unlike the isotope records, the sedimentary records are too coarse in their resolution to register quasi-cyclic activity at decadal to centennial scales. These higher resolution records therefore offer a means for possibly differentiating between the natural variability of tropical cyclone behaviour and the effects of anthropogenically induced global climate change. Recent studies with stalagmites in Belize shows that events can be determined on a week-by-week basis.

Historical Records

The *Royal Charter* which sank in a storm off Anglesey in 1859.

Before the invention of the telegraph in the early to mid-19th century, news was as fast as the fastest horse or stagecoach or ship. Normally, there was no advance warning of a tropical cyclone impact. However, the situation changed in the 19th century as seafaring people and land-based researchers, such as Father Viñes in Cuba, came up with systematic methods of reading the sky's appearance or the sea state, which could foretell a tropical cyclone's approach up to a couple days in advance.

One of the best documented storms is the Royal Charter Storm of 1859 which caused over 800 deaths in the UK alone. It led directly to the formation of the Meteorological Office under Robert Fitzroy. However, wind speed could not be measured accurately at the time, methods only becoming available after the Tay Bridge disaster of 1879. One of the better sources for storms in and around Britain is the shipwreck statistics compiled annually by the Board of Trade, but which have yet to be analysed in detail.

Michael Chenoweth used 18th century journals to reconstruct the climate of Jamaica. Together with Dmitry Divine, he also created a 318-year (1690–2007) record of tropical cyclones in the

Lesser Antilles, using newspaper accounts, ships' logbooks, meteorological journals, and other document sources.

In China, the abundance of historical documentary records in the form of *Fang Zhi* (semiofficial local gazettes) offers an extraordinary opportunity for providing a high-resolution historical dataset for the frequency of typhoon strikes.

Historical Climatology

Historical climatology is the study of historical changes in climate and their effect on human history and development. This differs from paleoclimatology which encompasses climate change over the entire history of Earth. The study seeks to define periods in human history where temperature or precipitation varied from what is observed in the present day. The primary sources include written records such as sagas, chronicles, maps and local history literature as well as pictorial representations such as paintings, drawings and even rock art. The archaeological record is equally important in establishing evidence of settlement, water and land usage.

Techniques of Historical Climatology

In literate societies, historians may find written evidence of climatic variations over hundreds or thousands of years, such as phenological records of natural processes, for example viticultural records of grape harvest dates. In preliterate or non-literate societies, researchers must rely on other techniques to find evidence of historical climate differences.

Past population levels and habitable ranges of humans or plants and animals may be used to find evidence of past differences in climate for the region. Palynology, the study of pollens, can show not only the range of plants and to reconstruct possible ecology, but to estimate the amount of precipitation in a given time period, based on the abundance of pollen in that layer of sediment or ice.

Evidence of Climatic Variations

The eruption of the Toba supervolcano, 70,000 to 75,000 years ago reduced the average global temperature by 5 degrees Celsius for several years and may have triggered an ice age. It has been postulated that this created a bottleneck in human evolution. A much smaller but similar effect occurred after the eruption of Krakatoa in 1883, when global temperatures fell for about 5 years in a row.

Before the retreat of glaciers at the start of the Holocene (~9600 BC), ice sheets covered much of the northern latitudes and sea levels were much lower than they are today. The start of our present interglacial period appears to have helped spur the development of human civilization.

Human Record

Evidence of a warm climate in Europe, for example, comes from archaeological studies of settlement and farming in the Early Bronze Age at altitudes now beyond cultivation, such as Dartmoor,

Exmoor, the Lake district and the Pennines in Great Britain. The climate appears to have deteriorated towards the Late Bronze Age however. Settlements and field boundaries have been found at high altitude in these areas, which are now wild and uninhabitable. Grimspound on Dartmoor is well preserved and shows the standing remains of an extensive settlement in a now inhospitable environment.

The 16th-century Skálholt Map of Norse America

Some parts of the present Saharan desert may have been populated when the climate was cooler and wetter, judging by cave art and other signs of settlement in Prehistoric Central North Africa.

One of Grimspound's hut circles

The Medieval Warm Period was a time of warm weather between about AD 800–1300, during the European Medieval period. Archaeological evidence supports studies of the Norse sagas which describe the settlement of Greenland in the 9th century AD of land now quite unsuitable for cultivation. For example, excavations at one settlement site have shown the presence of birch trees during the early Viking period. The same period records the discovery of an area called Vinland, probably in North America, which may also have been warmer than at present, judging by the alleged presence of grape vines. The interlude is known as the Medieval Warm Period.

Little Ice Age

Later examples include the Little Ice Age, well documented by paintings, documents (such as diaries) and events such as the River Thames frost fairs held on frozen lakes and rivers in the 17th and 18th centuries. The River Thames was made more narrow and flowed faster after old London Bridge was demolished in 1831, and the river was embanked in stages during the 19th century, both of which made the river less liable to freezing. Among the earliest references to the coming climate change is an entry in the *Anglo-Saxon Chronicle* dated 1046:

- "And in this same year after the 2nd of February came the severe winter with frost and snow, and with all kinds of bad weather, so that there was no man alive who could remember so severe a winter as that, both through mortality of men and disease of cattle; both birds and fishes perished through the great cold and hunger."

The *Chronicle* is the single most important historical source for the period in England between the departure of the Romans and the decades following the Norman Conquest. Much of the information given in the *Chronicle* is not recorded elsewhere.

The Frozen Thames, 1677

The Little Ice Age brought colder winters to parts of Europe and North America. In the mid-17th century, glaciers in the Swiss Alps advanced, gradually engulfing farms and crushing entire villages. The River Thames and the canals and rivers of the Netherlands often froze over during the winter, and people skated and even held frost fairs on the ice. The first Thames frost fair was in 1607; the last in 1814, although changes to the bridges and the addition of an embankment affected the river flow and depth, diminishing the possibility of freezes. The freeze of the Golden Horn and the southern section of the Bosphorus took place in 1622. In 1658, a Swedish army marched across the Great Belt to Denmark to invade Copenhagen. The Baltic Sea froze over, enabling sledge rides from Poland to Sweden, with seasonal inns built on the way. The winter of 1794/1795 was particularly harsh when the French invasion army under Pichegru could march on the frozen rivers of the Netherlands, while the Dutch fleet was fixed in the ice in Den Helder harbour. In the winter of 1780, New York Harbour froze, allowing people to walk from Manhattan to Staten Island. Sea ice surrounding Iceland extended for miles in every direction, closing that island's harbours to shipping.

The last written records of the Norse Greenlanders are from a 1408 marriage in Hvalsey Church — today the best-preserved of the Norse ruins

The severe winters affected human life in ways large and small. The population of Iceland fell by half, but this was perhaps also due to fluorosis caused by the eruption of the volcano Laki in 1783. Iceland also suffered failures of cereal crops and people moved away from a grain-based diet. The Norse colonies in Greenland starved and vanished (by the 15th century) as crops failed and livestock could not be maintained through increasingly harsh winters, though Jared Diamond noted that they had exceeded the agricultural carrying capacity before then. In North America, American Indians formed leagues in response to food shortages. In Southern Europe, in Portugal, snow storms were much more frequent while today they are rare. There are reports of heavy snowfalls in the winters of 1665, 1744 and 1886.

In contrast to its uncertain beginning, there is a consensus that the Little Ice Age ended in the mid-19th century.

Evidence of Anthropogenic Climate Change

Through deforestation and agriculture, some scientists have proposed a human component in some historical climatic changes. Human-started fires have been implicated in the transformation of much of Australia from grassland to desert. If true, this would show that non-industrialized societies could have a role in influencing regional climate. Deforestation, desertification and the salinization of soils may have contributed to or caused other climatic changes throughout human history.

Tornado Climatology

Tornadoes have been recorded on all continents except Antarctica and are most common in the middle latitudes where conditions are often favorable for convective storm development. The United States has the most tornadoes of any country, as well as the strongest and most violent tornadoes. A large portion of these tornadoes form in an area of the central United States popularly known as Tornado Alley. Other areas of the world that have frequent tornadoes include significant portions of Europe, South Africa, Bangladesh, parts of Argentina, Uruguay, and southern Brazil, New Zealand, and far eastern Asia.

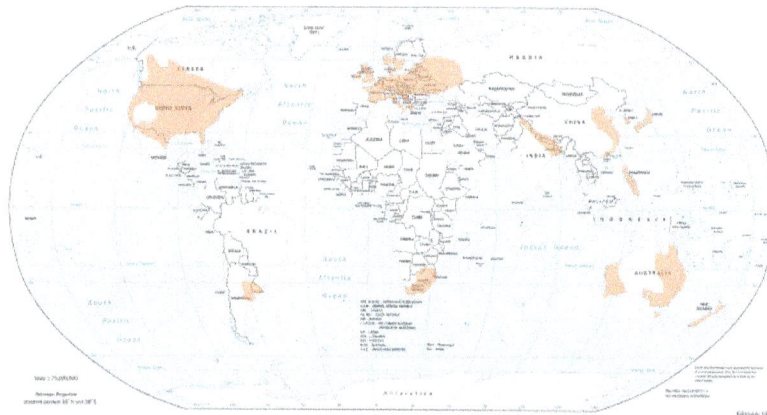

Areas worldwide with the highest frequency of tornadoes are indicated by orange shading.

The United States averaged 1,274 tornadoes per year in the last decade while Canada reports nearly 100 annually (largely in the southern regions). However, the UK has most tornadoes per area per year, 0.14 per 1000 km², although these tornadoes are generally weak, and many other European countries have a similar number of tornadoes per area.

The severity of tornadoes is commonly measured by the Enhanced Fujita Scale, which scales tornado intensity from EF0 to EF5 by wind speed and the amount of damage they do to human environments. These judgments are made after the tornado has dissipated and the damage trail is carefully studied by weather professionals.

Tornadoes are most common in spring and least common in winter. The seasonal transition during autumn and spring promotes the development of extratropical cyclones and frontal systems that support strong convective storms. Tornadoes are also common in landfalling tropical cyclones, where they are focused in the right poleward section of the cyclone. Tornadoes can also be spawned as a result of eyewall mesovortices, which persist until landfall. However, favorable conditions for tornado development can occur any time of the year.

Tornado occurrence is highly dependent on the time of day, because of solar heating. Worldwide, most tornadoes occur in the late afternoon, between 3 pm and 7 pm local time, with a peak near 5 pm. Destructive tornadoes can occur at any time of day, as evidenced by the Gainesville Tornado of 1936 (one of the deadliest tornadoes in history) that occurred at 8:30 am local time.

Geography

The United States has the most tornadoes of any country. Many of these form in an area of the central (with some definitions including Southern) United States known as Tornado Alley. This area extends into Canada, particularly the Prairie Provinces and Ontario; however, activity in Canada is less frequent and intense than that of the US. The high frequency of tornadoes in North America is largely due to geography, as moisture from the Gulf of Mexico is easily advected into the mid-continent with few topographic barriers in the way. The Rocky Mountains block Pacific-sourced moisture and buckle the atmospheric flow, forcing drier air at mid-levels of the troposphere due to downsloping winds and causing cyclogenesis downstream to the east of the mountains. Downsloping winds off the Rockies force the formation of a dry line when the flow aloft is strong, while the Gulf of Mexico fuels abundant low-level moisture. This unique topography allows for frequent collisions of warm and cold air, the conditions that breed strong, long-lived storms throughout the year. This area extends into Canada, particularly Ontario and the Prairie Provinces, and strong tornadoes can also occur in northern Mexico.

A large region of South America is characterized by storms that reach the level of supercells and produce intense hailstorms, floods, and tornadoes during the spring, summer, and early fall. The region recently appointed as the Tornado Corridor (South America) is considered as the second largest in the world in terms of the formation of extreme weather events. It covers most of central Argentina, southern Paraguay, southeastern Brazil, and Uruguay.

Bangladesh and surrounding areas of eastern India suffer from a couple tornadoes annually of similar severity to stronger tornadoes in the US. These occur with a greater recurrence interval (although over a smaller region), and tend to be under-reported due to the scarcity of media coverage of a developing country. The annual human death toll from tornadoes in Bangladesh is estimated at about 179 deaths per year, which is much greater than in the US. This is likely due to the density of population, poor quality of construction, lack of tornado safety knowledge and warnings, and other factors.

Other areas of the world that have frequent strong tornadoes include Germany, the Czech Republic, Slovakia, Italy, Spain, China, and the Philippines. Australia, France, Russia, areas of the Middle East, Japan, and parts of Mexico have a history of multiple damaging tornado events.

Tornadoes in the USA

The United States averaged 1,274 tornadoes per year in the last decade. April 2011 saw the most tornadoes ever recorded for any month in the US National Weather Service's history, 875; the previous record was 542 in one month. It has more tornadoes yearly than any other country and reports more violent (F4 and F5) tornadoes than anywhere else.

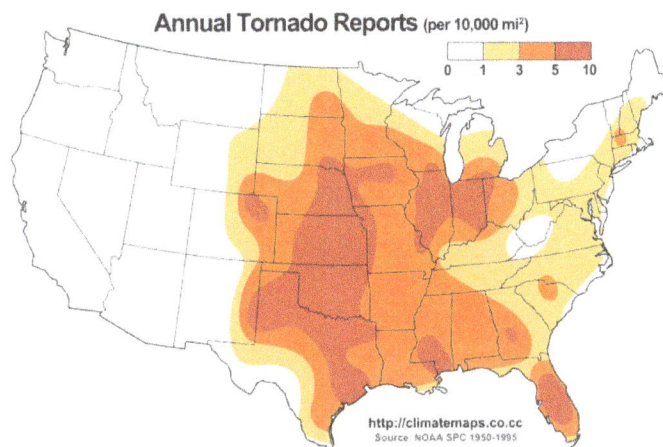

Average annual tornado reports in the United States.

Tornadoes are common in many states but are most common to the west of the Appalachian Mountains and to the east of the Rockies. The Atlantic seaboard states – North Carolina, South Carolina, Georgia and Virginia – are also very vulnerable, as well as Florida. The areas most vulnerable to tornadoes are the Southern Plains and Florida, though most Florida tornadoes are relatively weak. The Southern USA is one of the worst affected regions in terms of casualties.

Tornado reports have been officially collated since 1950. These reports have been gathered by the National Climatic Data Center (NCDC), based in Asheville, North Carolina. A tornado can be reported more than once, such as when a storm crosses a county line and reports are made from two counties.

Common Misconceptions

Some people mistakenly believe that tornadoes only occur in the countryside. This is hardly the case. While it is true that the plains states are the most tornado-prone places in the nation, it should be noted that tornadoes have been reported in every U.S state, including Alaska and Hawaii. One likely reason why tornadoes are so common in the central U.S is because this is where Arctic air first collides with warm tropical air from the Gulf of Mexico where the cold front has not been "weakened" yet. As it heads further east, however, it is possible for the front to lose its strength as it travels over more warm air. Therefore, tornadoes are not as common on the East Coast as they are in the Midwest. However, they have happened on rare occasion, such as the F3 twister that struck Limerick Township, Pennsylvania near Philadelphia on 27 July 1994; the F2 twister that struck the northern suburbs of New York City on 12 July 2006; the EF2 twister in the borough of Brooklyn on 8 August 2007; or the F4 twister that struck La Plata, Maryland on 28 April 2002.

Tornadoes can occur west of the continental divide, but they are infrequent and usually relatively weak and short-lived. Recently tornadoes have struck the Pacific coast town of Lincoln City, Oregon (1996); Sunnyvale, California (1998); and downtown Salt Lake City, Utah (1999). The California Central Valley is an area of some frequency for tornadoes, albeit of weak intensity. Though tornadoes that occur on the Western Seaboard typically are weak, more powerful, damaging tornadoes can occur, such as a tornado that occurred on 22 May 2008 in Per-ris, California.

More tornadoes occur in Texas than in any other US state. The state which has the highest number of tornadoes per unit area is Florida, although most of the tornadoes in Florida are weak tornadoes of EF0 or EF1 intensity. A number of Florida's tornadoes occur along the edge of hurricanes that strike the state. The state with the highest number of strong tornadoes per unit area is Oklahoma. The neighboring state of Kansas is also a particularly notorious tornado state. It records the most EF4 and EF5 tornadoes in the country.

Tornadoes in Canada

Canada also experiences numerous tornadoes, although fewer than the United States. In Canada, at least 80–100 tornadoes occur annually (with many more likely undetected in large expanses of unpopulated areas), causing tens of millions of dollars in damage. Most are weak F0 or F1 in intensity, but there are on average a few F2 or stronger that touch down each season.

For example, the tornado frequency of Southwestern Ontario is about half that of the most prone areas of the central US plains. The last multiple tornado-related deaths in Canada were caused by a tornado in Ear Falls, Ontario, on 9 July 2009, where 3 died, and the last killer tornado was on 21 August 2011 in Goderich, Ontario. The two deadliest tornadoes on Canadian soil were the Regina Cyclone of 30 June 1912 (28 fatalities) and the Edmonton Tornado of 31 July 1987 (27 fatalities). Both of these storms were rated an F4 on the Fujita scale. The city of Windsor was struck by strong tornadoes four times within a 61-year span (1946, 1953, 1974, 1997) ranging in strength from an F2 to F4. Windsor has been struck by more significant tornadoes than any other city in Canada. Canada's first official F5 tornado struck Elie, Manitoba on 22 June 2007. At least two other tornadoes in Saskatchewan in earlier parts of the 20th century are suspected as F5. Tornadoes are most frequent in the provinces of Alberta, Saskatchewan, Manitoba, and Ontario.

Europe

Europe has about 300 tornadoes per year – much more than estimated by Alfred Wegener in his classic book *Wind- und Wasserhosen in Europa* ("Tornadoes and Waterspouts in Europe"). They are most common in June–August, especially in the inlands – rarest in January–March. Strong and violent tornadoes (F3–F5) do occur, especially in some of the interior areas and in the south – but are not as common as in parts of the USA. As in the USA, tornadoes are far from evenly distributed. Europe has some small "tornado alleys" – probably because of frontal collisions as in the south and east of England, but also because Europe is partitioned by mountain ranges like the Alps. Parts of Styria (Steiermark) in Austria may be such a tornado alley, and this county has had at least three F3 tornadoes since 1900. F3 and perhaps one F4 tornado have occurred as far north as Finland.

Since 1900, deadly tornadoes have occurred in Austria, Belgium, Cyprus, England, Finland, France, Germany, Italy (such as the F5/T10 of Udine-Treviso on 24 July 1930, which killed 23 people,), Malta, Poland, Portugal (such as the F3/T7 of Castelo Branco on 6 November 1954, which killed 5 and injured 220), Romania, Russia, Scotland, and Wales. The 1984 Ivanovo–Yaroslavl outbreak, with more than 400 fatalities and 213 injured, was the century's deadliest tornado or outbreak in Europe. It included at least one F5 and one F4. Europe's perhaps deadliest tornado ever (and probably one of the World's deadliest tornadoes) hit Malta in 1551 (or 1556) and killed about 600.

One notable tornado of recent years was the Birmingham Tornado (UK) which struck Birmingham, United Kingdom, in July 2005. A row of houses was destroyed, but no one was killed. A strong F3 (T7) Tornado hit the small town Micheln in Saxony-Anhalt, Germany on 23 July 2004 leaving 6 people injured and more than 250 buildings massively damaged.

Asia

Bangladesh and the eastern parts of India are very exposed to destructive tornadoes causing higher deaths and injuries. Bangladesh, Philippines, and Japan has the highest number of reported tornadoes in Asia.

In June of 2016, a storm producing multiple tornadoes and hail struck a densely populated area of farms and factories near the city of Yancheng in Jiangsu province, about 800 kilometers (500 miles) south of Beijing, killing at least 78 people and destroying buildings. Nearly 500 people were injured, 200 of them critically, the official Xinhua News Agency reported.

South America

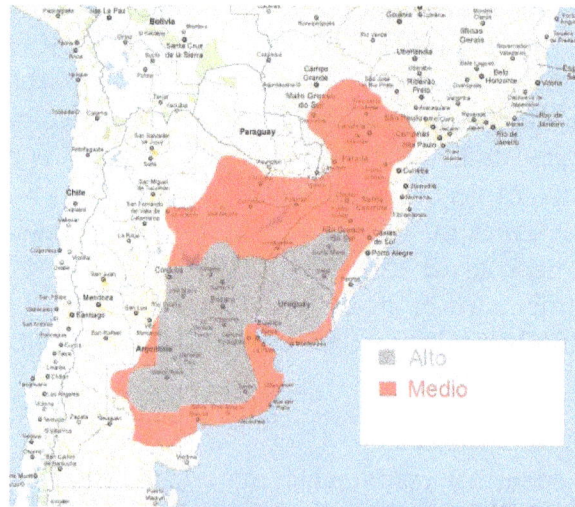

Zone of tornadoes (South America).Gray zone high risk, red zone medium risk

South America has the hall of tornadoes composed by central and northern Argentina, southern Brazil, Uruguay, and part of Paraguay, considered the second highest frequency tornado region in the world. Argentina has pockets with high tornadic activity, and has the strongest tornadoes in the southern hemisphere like the F5 registered in San Justo, and the tornado outbreak in Buenos Aires with more than 300 tornadoes registered in less than 24 hours. Argentina has the second most tornadoes on the American continent and the world (after the United States) with approximately 300 tornadoes every year depending on the El Niño current, the La Niña current, or neutral year . This region is propitious for tornadoes and severe thunderstorms, due to the large size of the Pampas Plain where the cold air from Patagonia and Antarctica, collides with warm, moist air from areas of Brazil, region northern Argentina and Paraguay, this is added to the dry air from the Andes which collide in this region producing intense storms that frequently producing tornadic supercells. A study of NASA found to the strongest storm occur in the east of the Andes Mountains in Argentina

South American Tornados

On 16 September 1816, one of the first tornados recorded destroyed the town of Rojas (240 km west of the city of Buenos Aires) On 20 September 1926, an EF4 tornado struck the city of Encarnación (Paraguay) and killed over 300 people, making it the second deadliest tornado in South America. On 21 April 1970, an F4 struck Fray Marcos in Uruguay and killed 11, making it the strongest in Uruguay's history. On 10 January 1973, an F5 struck the city of San Justo, Argentina, 105 km north of the city of Santa Fe. The San Justo tornado is considered the worst tornado ever to occur in the Southern Hemisphere, with winds that exceeded 400 km/h. On 6 May 1992, the people of López Station, Argentina were devastated by an EF4 tornado that caused 4 deaths among the 150 residents. On 13 April 1993, the province of Buenos Aires (Argentina) was impacted by the largest tornado outbreak in South American history. More than 300 tornadoes were recorded over 24 hours with intensities ranging from F1 to F3. On 28 October 1978, an EF4 tornado with winds of 270 km/h hit the city of Mortars in the province of Córdoba, killing 5 people. On 26 December 2003 the 2003 Córdoba Tornado struck 6 km west of Córdoba's city center. It was rated an F3 with winds exceeding 300 km/h, killing 5 and injuring hundreds. On the night of 7 September 2009, an EF4 tornado destroyed the town of San Pedro, Argentina (Misiones), killing 11 people. The same tornado hit the nearby town of Guaraciaba, Brazil, killing 6. The neighboring towns Veloso Santo and Santa Cecilia were seriously damaged and were declared in a state of emergency.

Africa

Tornadoes do occur in South Africa. In October 2011 (i.e. in the spring), two people were killed and nearly 200 were injured after a tornado formed, near Ficksburg in the Free State; more than 1,000 shacks and houses were flattened.

Oceania

Australia has about 16 tornadoes per year – excluding waterspouts, which are common. In New Zealand, a tornado hit the northern suburbs of Auckland on 3 May 2011, killing one and injuring at least 16 people.

Frequency of Occurrence

Tornadoes can form in any month, providing the conditions are favorable. For example, a freak tornado hit South St. Louis County Missouri on 31 December 2010, causing pockets of heavy damage to a modest area before dissipating. The temperature was unseasonably warm that day. They are least common during the winter and most common in spring. Since autumn and spring are transitional periods (warm to cool and vice versa) there are more chances of cooler air meeting with warmer air, resulting in thunderstorms. Tornadoes in the late summer and fall can also be caused by hurricane landfall.

Not every thunderstorm, supercell, squall line, or tropical cyclone will produce a tornado. Precisely the right atmospheric conditions are required for the formation of even a weak tornado. On the other hand, 700 or more tornadoes a year are reported in the contiguous United States.

On average, the United States experiences 100,000 thunderstorms each year, resulting in more than 1,200 tornadoes and approximately 50 deaths per year. The deadliest U.S. tornado recorded is the 18 March 1925, Tri-State Tornado that swept across southeastern Missouri, southern Illinois and southern Indiana, killing 695 people. The biggest tornado outbreak on record—with 353 tornadoes over the course of just 3 1/2 days, including four F5 and eleven F4 tornadoes—occurred starting on 25 April 2011 and intensifying on 26 April and especially the record-breaking day of 27 April before ending on 28 April. It is referred to as the 2011 Super Outbreak. Previously, the record was 148 tornadoes, dubbed the 1974 Super Outbreak. Another such significant storm system was the Palm Sunday tornado outbreak of 1965, which affected the United States Midwest on 11 April 1965. A series of continuous tornado outbreaks is known as a tornado outbreak sequence, with significant occurrences in May 1917, 1930, 1949, and 2003.

Time of Occurrence

Diurnality

Tornado occurrence is highly dependent on the time of day. Austria, Finland, Germany, and the

United States' peak hour of occurrence is 5 pm, with roughly half of all tornado occurrence between 3 p.m. and 7 p.m. local time, due to this being the time of peak atmospheric heating, and thus the maximum available energy for storms; some researchers, including Howard B. Bluestein of the University of Oklahoma, have referred to this phenomenon as "five o'clock magic." Despite this, there are several morning tornadoes reported, like the Seymour, Texas one in April 1980. The most recent happened on 25 May 2015 an F3 tornado raged through Acuna City a Mexico border city at 5:45 am. Killing 14 people.

Seasonality

The time of year is a big factor of the intensity and frequency of tornadoes. On average, in the United States as a whole, the month with the most tornadoes is May, followed by the months June, April, and July. There is no "tornado season" though, as tornadoes, including violent tornadoes and major outbreaks, can and do occur anywhere at any time of year if favorable conditions develop. Major tornado outbreaks have occurred in every month of the year.

July is the peak month in Austria, Finland, and Germany. On average, there are around 294 tornadoes throughout the United States during the month of May, and as many as 543 tornadoes have been reported in the month of May alone (in 2003). The months with the fewest tornadoes are usually December and January, although major tornado outbreaks can and sometimes do occur even in those months. In general, in the Midwestern and Plains states, springtime (especially the month of May) is the most active season for tornadoes, while in the far northern states (like Minnesota and Wisconsin), the peak tornado season is usually in the summer months (June and July). In the colder late autumn and winter months (from early December to late February), tornado activity is generally limited to the southern states, where it is possible for warm Gulf of Mexico air to penetrate.

The reason for the peak period for tornado formation being in the spring has much to do with temperature patterns in the U.S. Tornadoes often form when cool, polar air traveling southeastward from the Rockies overrides warm, moist, unstable Gulf of Mexico air in the eastern states. Tornadoes therefore tend to be commonly found in front of a cold front, along with heavy rains, hail, and damaging winds. Since both warm and cold weather are common during the springtime, the conflict between these two air masses tends to be most common in the spring. As the weather warms across the country, the occurrence of tornadoes spreads northward. Tornadoes are also common in the summer and early fall because they can also be triggered by hurricanes, although the tornadoes caused by hurricanes are often much weaker and harder to spot. Winter is the least common time for tornadoes to occur, since hurricane activity is virtually non-existent at this time, and it is more difficult for warm, moist maritime tropical air to take over the frigid Arctic air from Canada, occurrences are found mostly in the Gulf states and Florida during winter (although there have been some notable exceptions). Interestingly, there is a second active tornado season of the year, late October to mid-November. Autumn, like spring, is a time of the year when warm weather alternates with cold weather frequently, especially in the Midwest, but the season is not as active as it is during the springtime and tornado frequencies are higher along the Atlantic Coastal plain as opposed to the Midwest. They usually appear in late summer.

Long-term Trends

The reliable climatology of tornadoes is limited in geographic and temporal scope; only since

1976 in the United States and 2000 in Europe have thorough and accurate tornado statistics been logged. However, some trends can be noted in tornadoes causing significant damage in the United States, as somewhat reliable statistics on damaging tornadoes exist as far back as 1880. The highest incidence of violent tornadoes seems to shift from the Southeastern United States to the southern Great Plains every few decades. Also, the 1980s seemed to be a period of unusually low tornado activity in the United States, and the number of multi-death tornadoes decreased every decade from the 1920s to the 1980s, suggesting a multi-decadal pattern of some sort.

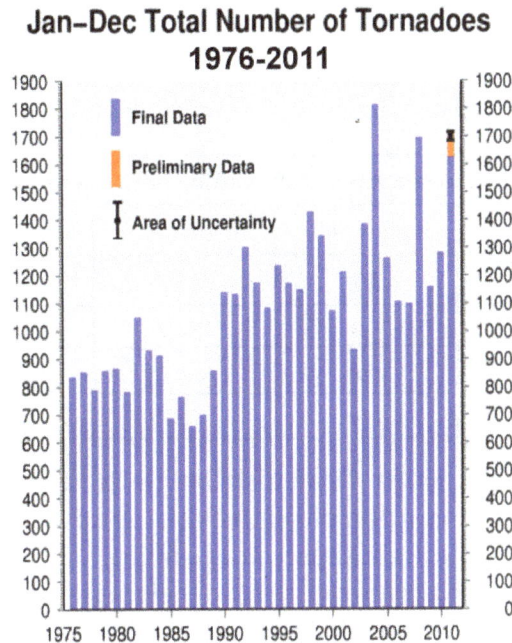

U. S. Annual January – December Tornado Count 1976–2011 from NOAA National Climatic Data Center

Tropical Cyclone Rainfall Climatology

A tropical cyclone rainfall climatology is developed to determine rainfall characteristics of past tropical cyclones. A tropical cyclone rainfall climatology can be used to help forecast current or upcoming tropical cyclone impacts. The degree of a tropical cyclone rainfall impact depends upon speed of movement, storm size, and degree of vertical wind shear. One of the most significant threats from tropical cyclones is heavy rainfall. Large, slow moving, and non-sheared tropical cyclones produce the heaviest rains. The intensity of a tropical cyclone appears to have little bearing on its potential for rainfall over land, but satellite measurements over the last several years show that more intense tropical cyclones produce noticeably more rainfall over water. Flooding from tropical cyclones remains a significant cause of fatalities, particularly in low-lying areas.

Anticipating a Flood Event

While inland flooding is common to tropical cyclones, there are factors which lead to excessive rainfall from tropical cyclones. Slow motion, as was seen during Hurricane Danny (1997) and Hurricane Wilma, can lead to high amounts of rainfall. The presence of mountains/hills near the coast, like across much of Mexico, Haiti, the Dominican Republic, Central America, Madagascar, Réunion, China, and Japan acts to magnify rainfall potential due to forced upslope flow into the mountains. Strong upper level forcing from a trough moving through the Westerlies and its associated cold front, as was the case during Hurricane Floyd, can lead to high amounts even from systems moving at an average forward motion. Larger tropical cyclones drop more rainfall as they precipitate upon one spot for a longer time frame than average or small tropical cyclones. A combination of two of these factors could be especially crippling, as was seen during Hurricane Mitch in Central America. During the 2005 season, flooding related to slow-moving Hurricane Stan's broad circulation led to 1,662–2,000 deaths.

General Distribution within a Tropical Cyclone

Rainfall Rate per day within radius of the center (Riehl)			
Radius (mi)	Radius (km)	Amount (in)	Amount (mm)
35	56	33.98	863
70	112	13.27	337
140	224	4.25	108
280	448	1.18	30

Isaac Cline was the first to investigate rainfall distribution around tropical cyclones in the early 1900s. He found that a larger proportion of rainfall falls in advance of the center (or eye) than after the center's passage, with the highest percentage falling in the right front quadrant. Father Viñes of Cuba found that some tropical cyclones have their highest rainfall rates in the rear quadrant within a training (non-moving) inflow band. Normally, as a tropical cyclone intensifies, its heavier rainfall rates become more concentrated around its center. Rainfall is found to be heaviest in tropical cyclone's inner core, whether it be the eyewall or central dense overcast, within a degree latitude of the center, with lesser amounts farther away from the center. Most of the rainfall in tropical cyclones is concentrated within its radius of gale-force (34 knots/39 mph/63 km/h) winds. Rainfall is more common near the center of tropical cyclones overnight. Over land, outer bands are more active during the heating of the day, which can act to restrict inflow into the center of the cyclone. Recent studies have shown that half of the rainfall within a tropical cyclone is stratiform in nature. The chart to the right was developed by Riehl in 1954 using meteorological equations that assume a gale radius of about 140 miles (230 km), a fairly symmetric cyclone, and does not consider topographic effects or vertical wind shear. Local amounts can exceed this chart by a factor of two due to topography. Wind shear tends to lessen the amounts below what is shown on the table.

Relation to Storm Size

Larger tropical cyclones have larger rain shields, which can lead to higher rainfall amounts farther from the cyclone's center. This is generally due to the longer time frame rainfall falls at any one spot in a larger system, when compared to a smaller system. Some of the difference seen concerning rainfall between larger and small storms could be the increased sampling of rainfall within a larger tropical cyclone when compared to that of a compact cyclone; in other words, the difference could be the result of a statistical problem.

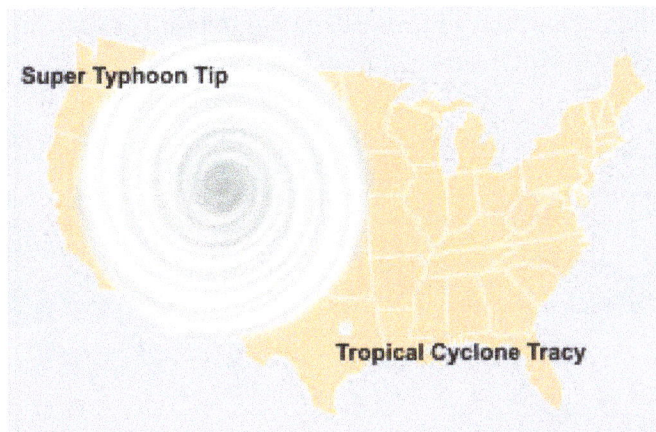

The relative sizes of Typhoon Tip, Cyclone Tracy, and the United States.

Slow/Looping Motion on Rainfall Magnitude

Storms which have moved slowly, or loop, over a succession of days lead to the highest rainfall amounts for several countries. Riehl calculated that 33.97 inches (863 mm) of rainfall per day can be expected within one-half degree, or 35 miles (56 km), of the center of a mature tropical cyclone. Many tropical cyclones progress at a forward motion of 10 knots, which would limit the duration of this excessive rainfall to around one-quarter of a day, which would yield about 8.50 inches (216 mm) of rainfall. This would be true over water, within 100 miles (160 km) of the coastline, and outside topographic features. As a cyclone moves farther inland and is cut off from its supply of warmth and moisture (the ocean), rainfall amounts from tropical cyclones and their remains decrease quickly.

Vertical Wind Shear Impact on Rainfall Shield

Vertical wind shear forces the rainfall pattern around a tropical cyclone to become highly asymmetric, with most of the precipitation falling to the left and downwind of the shear vector, or downshear left. In other words, southwesterly shear forces the bulk of the rainfall north-northeast of the center. If the wind shear is strong enough, the bulk of the rainfall will move away from the center leading to what is known as an exposed circulation center. When this occurs, the potential magnitude of rainfall with the tropical cyclone will be significantly reduced.

Effect of Interaction with Frontal Boundaries/Upper Level Troughs

As a tropical cyclone interacts with an upper-level trough and the related surface front, a distinct northern area of precipitation is seen along the front ahead of the axis of the upper level trough. This type of interaction can lead to the appearance of the heaviest rainfall falling along and to the left of the tropical cyclone track, with the precipitation streaking hundreds of miles or kilometers downwind from the tropical cyclone. The stronger the upper trough picking up the tropical cyclone, the more significant the left of track shift in the rainfall distribution tends to be.

Mountains

Moist air forced up the slopes of coastal hills and mountain chains can lead to much heavier rainfall than in the coastal plain. This heavy rainfall can lead to landslides, which still cause significant loss of life such as seen during Hurricane Mitch in Central America.

Global Distribution

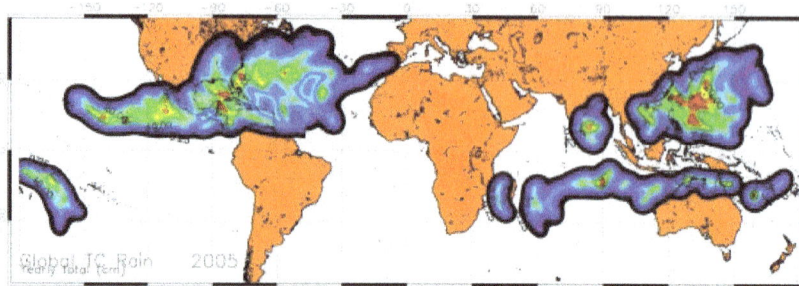

Global tropical cyclone rainfall in 2005

Globally, tropical cyclone rainfall is more common across the northern hemisphere than across the southern hemisphere. This is mainly due to the normal annual tropical cyclone distribution, as between half and two-thirds of all tropical cyclones form north of the equator. Rainfall is concentrated near the 15th parallel in both hemispheres, with a less steep dropoff seen with latitude across the northern hemisphere, due to the stronger warm water currents seen in that hemisphere which allow tropical cyclones to remain tropical in nature at higher latitudes than south of the equator. In the southern hemisphere, rainfall impacts will be most common between January and March, while north of the equator, tropical cyclone rainfall impacts are more common between June and November. Japan receives over half of its rainfall from typhoons.

United States Tropical Cyclone Rainfall Statistics

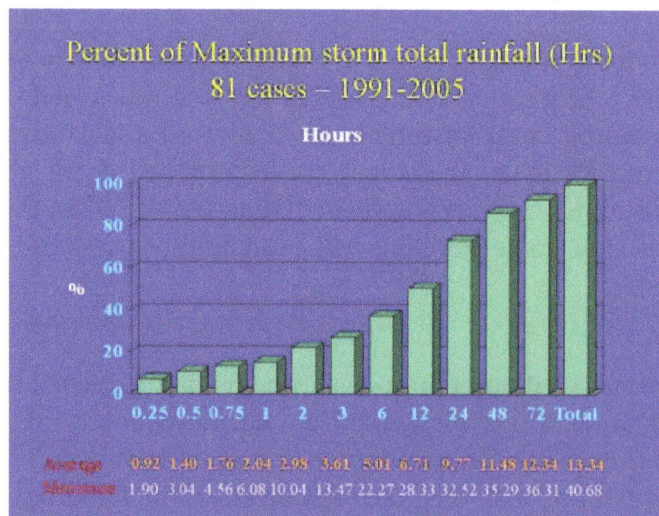

U.S. Tropical Cyclone Rainfall Accumulations per time frame

Between 1970-2004, inland flooding from tropical cyclones caused a majority of the fatalities in the United States. This statistic changed in 2005, when Hurricane Katrina's impact alone shifted the most deadly aspect of tropical cyclones back to storm surge, which has historically been the most deadly aspect of strong tropical cyclones. On average, five tropical cyclones of at least tropical depression strength lead to rainfall across the contiguous United States annually, contributing around a quarter of the annual rainfall to the southeast United States. While many of these storms form in the Atlantic Basin, some systems or their remnants move through Mexico from the Eastern Pacific Basin. The average storm total rainfall for a tropical cyclone impacting the lower 48 from the Atlantic Basin is about 16 inches (406 mm), with 70–75 percent of the storm total falling within a 24-hour period. The highest point total was seen during Amelia 1978, when 48 inches (1,218 mm) fell upon central Texas.

Urban Climatology

Urban climatology refers to a specific branch of climatology that is concerned with interactions between urban areas and the atmosphere, the effects they have on one another, and the varying spatial and temporal scales at which these processes (and responses) occur.

History

Luke Howard is considered to have established urban climatology with his book *The Climate of London*, which contained continuous daily observations from 1801 to 1841 of wind direction, atmospheric pressure, maximum temperature, and rainfall.

Urban climatology came about as a methodology for studying the results of industrialization and urbanization. Constructing cities changes the physical environment and alters energy, moisture, and motion regimes near the surface. Most of these alterations can be traced to causal factors such as air pollution; anthropogenic sources of heat; surface waterproofing; thermal properties of the surface materials; and morphology of the surface and its specific three-dimensional geometry—building spacing, height, orientation, vegetative layering, and the overall dimensions and geography of these elements. Other factors are relief, proximity to water bodies, size of the city, population density, and land-use distributions.

Influential Factors

Several factors influence the urban climate, including city size, the morphology of the city, land-use configuration, and the geographic setting (such as relief, elevation, and regional climate). Some of the differences between urban and rural climates include air quality, wind patterns, and changes in rainfall patterns, but one of the most studied is the urban heat island effect.

Temperature and Urban Heat Island Effect

Urban environments are typically warmer than their surroundings, as documented over a century ago by Howard. Urban areas are islands or spots on the broader scale compared with more rural surrounding land. The spatial distribution of temperatures occurs in tandem with temporal changes, which are both causally related to anthropogenic sources.

The urban environment has two atmosphere layers, besides the planetary boundary layer outside and extending well above the city: (1) The urban boundary layer is due to the spatially integrated heat and moisture exchanges between the city and its overlying air. (2) The surface of the city corresponds to the level of the urban canopy layer. Fluxes across this plane comprise those from individual units, such as roofs, canyon tops, trees, lawns, and roads, integrated over larger land-use divisions (for example, suburbs). The urban heat island effect has been a major focus of urban climatological studies, and in general the effect the urban environment has on local meteorological conditions.

Pollution

The field also includes the topics of air quality, Radiation Fluxes, Micro-Climates and even issues traditionally associated with architectural design and engineering, such as Wind Engineering. Causes and effects of pollution as understood through Urban Climatology are becoming more important for Urban Planning.

Precipitation

Changes in winds and convection patterns over and around cities impacts precipitation. Contributing factors are believed to be urban heat island, heightened surface roughness, and increased aerosol concentration.

Climate Change

Urban climatology is strongly linked to research surrounding global warming. As centers for socio-economic activities, cities produce large amounts of greenhouse gases, most notably CO_2 as a consequence of human activities such as transport, development, waste related to heating and cooling requirements etc.

Globally, cities are expected to grow into the 21st century (and beyond) - as they grow and develop the landscapes in which they inhabit will change so too will the atmosphere resting above them, increasing emissions of GHG's thus contributing to the global green house effect.

Finally, many cities are vulnerable to the projected consequences of climate change (sea level rise, changes in temperature, precipitation, storm frequency) as most develop on or near coast-lines, nearly all produce distinct urban heat islands and atmospheric pollution: as areas in which there is concentrated human habitation these effects potentially will have the largest and most dramatic impact (e.g. France's heat wave in 2003) and thus are a major focus for urban climatology.

Spatial Planning and Public Health

Urban Climatology impacts decision-making for municipal planning and policy in regards to pollution, extreme heat events, and stormwater modeling.

Dendroclimatology

Dendroclimatology is the science of determining past climates from trees (primarily properties of the annual tree rings). Tree rings are wider when conditions favor growth, narrower when times are difficult. Other properties of the annual rings, such as maximum latewood density (MXD) have been shown to be better proxies than simple ring width. Using tree rings, scientists have estimated many local climates for hundreds to thousands of years previous. By combining multiple tree-ring studies (sometimes with other climate proxy records), scientists have estimated past regional and global climates.

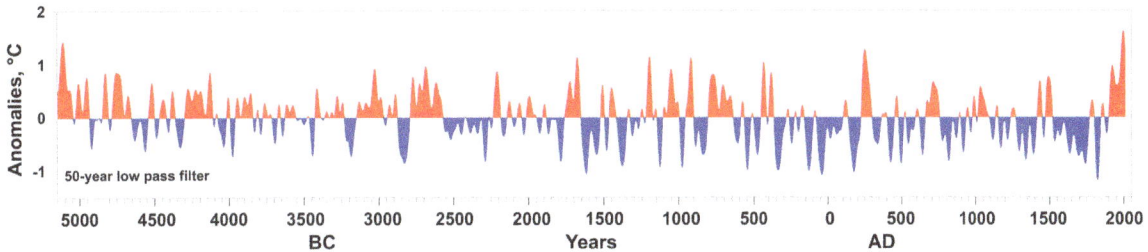

Variation of tree ring width translated into summer temperature anomalies for the past 7000 years, based on samples from holocene deposits on Yamal Peninsula and Siberian now living conifers.

Advantages

Tree rings are especially useful as climate proxies in that they can be well-dated (via matching of the rings from sample to sample, i.e. dendrochronology). This allows extension backwards in time using deceased tree samples, even using samples from buildings or from archeological digs. Another advantage of tree rings is that they are clearly demarked in annual increments, as opposed to other proxy methods such as boreholes. Furthermore, tree rings respond to multiple climatic effects (temperature, moisture, cloudiness), so that various aspects of climate (not just temperature) can be studied. However, this can be a double-edged sword as discussed in Climate factors.

Limitations

Along with the advantages of dendroclimatology are some limitations: confounding factors, geographic coverage, annular resolution, and collection difficulties. The field has developed various methods to partially adjust for these challenges.

Confounding Factors

There are multiple climate and non-climate factors as well as nonlinear effects that impact tree ring width. Methods to isolate single factors (of interest) include botanical studies to calibrate growth influences and sampling of "limiting stands" (those expected to respond mostly to the variable of interest).

Climate Factors

Climate factors that affect trees include temperature, precipitation, sunlight, and wind. To differentiate among these factors, scientists collect information from "limiting stands." An example of a limiting stand is the upper elevation treeline: here, trees are expected to be more affected by temperature variation (which is "limited") than precipitation variation (which is in excess). Conversely, lower elevation treelines are expected to be more affected by precipitation changes than temperature variation. This is not a perfect work-around as multiple factors still impact trees even at the "limiting stand," but it helps. In theory, collection of samples from nearby limiting stands of different types (e.g. upper and lower treelines on the same mountain) should allow mathematical solution for multiple climate factors. However, this method is rarely used.

Non-climate Factors

Non-climate factors include soil, tree age, fire, tree-to-tree competition, genetic differences, logging or other human disturbance, herbivore impact (particularly sheep grazing), pest outbreaks, disease, and CO_2 concentration. For factors which vary randomly over space (tree to tree or stand to stand), the best solution is to collect sufficient data (more samples) to compensate for confounding noise. Tree age is corrected for with various statistical methods: either fitting spline curves to the overall tree record or using similar aged trees for comparison over different periods (regional curve standardization). Careful examination and site selection helps to limit some confounding effects, for example picking sites undisturbed by modern man.

Non-linear Effects

In general, climatologists assume a linear dependence of ring width on the variable of interest (e.g. moisture). However, if the variable changes enough, response may level off or even turn opposite. The home gardener knows that one can underwater or overwater a house plant. In addition, it is possible that interaction effects may occur (for example "temperature times precipitation" may affect growth as well as temperature and precipitation on their own. Here, also, the "limiting stand" helps somewhat to isolate the variable of interest. For instance, at the upper treeline, where the tree is "cold limited", it's unlikely that nonlinear effects of high temperature ("inverted quadratic") will have numerically significant impact on ring width over the course of a growing season.

Botanical Inferences to Correct for Confounding Factors

Botanical studies can help to estimate the impact of confounding variables and in some cases guide corrections for them. These experiments may be either ones where growth variables are all controlled (e.g. in a greenhouse), partially controlled (e.g. FACE [Free Airborne Concentration Enhancement] experiments—add ref), or where conditions in nature are monitored. In any case, the important thing is that multiple growth factors are carefully recorded to determine what impacts growth. (Insert Fennoscandanavia paper reference). With this information, ring width response can be more accurately understood and inferences from historic (unmonitored) tree rings become more certain. In concept, this is like the limiting stand principle, but it is more quantitative—like a calibration.

Divergence Problem

The divergence problem is the disagreement between the temperatures measured by the thermometers (instrumental temperatures) on one side, and the temperatures reconstructed from the latewood density or width of tree rings on the other side, at many treeline sites in northern forests.

While the thermometer records indicate a substantial warming trend, tree rings from these particular sites do not display a corresponding change in their maximum latewood density or, in some cases, their width. This does not apply to all such studies. Where this applies, a temperature trend extracted from tree rings alone would not show any substantial warming. The temperature graphs calculated from instrumental temperatures and from these tree ring proxies thus "diverge" from one another since the 1950s, which is the origin of the term. This divergence raises obvious questions of whether other, unrecognized divergences have occurred in the past, prior to the era of thermometers. There is evidence suggesting that the divergence is caused by human activities, and so confined to the recent past, but use of affected proxies can lead to overestimation of past temperatures, understating the current warming trend. There is continuing research into explanations and ways to avoid this problem with tree ring proxies.

Geographic Coverage

Trees do not cover the Earth. Polar and marine climates cannot be estimated from tree rings. In perhumid tropical regions, Australia and southern Africa, trees generally grow all year round and don't show clear annual rings. In some forest areas, the tree growth is too much influenced by multiple factors (no "limiting stand") to allow clear climate reconstruction. The coverage difficulty is dealt with by acknowledging it and by using other proxies (e.g. ice cores, corals) in difficult areas. In some cases it can be shown that the parameter of interest (temperature, precipitation, etc.) varies similarly from area to area, for example by looking at patterns in the instrumental record. Then one is justified in extending the dendroclimatology inferences to areas where no suitable tree ring samples are obtainable.

Annular Resolution

Tree rings show the impact on growth over an entire growing season. Climate changes deep in the dormant season (winter) will not be recorded. In addition, different times of the growing season may be more important than others (i.e. May versus September) for ring width. However, in general the ring width is used to infer the overall climate change during the corresponding year (an approximation). Another problem is "memory" or autocorrelation. A stressed tree may take a year or two to recover from a hard season. This problem can be dealt with by more complex modeling (a "lag" term in the regression) or by reducing the skill estimates of chronologies.

Collection Difficulties

Tree rings must be obtained from nature, frequently from remote regions. This means that special efforts are needed to map sites properly. In addition, samples must be collected in difficult (often sloping terrain) conditions. Generally, tree rings are collected using a hand-held borer device, that requires skill to get a good sample. The best samples come from felling a tree and sectioning it. However, this requires more danger and does damage to the forest. It may not be allowed in certain areas, particularly with the oldest trees in undisturbed sites (which are the most interesting scientifically). As with all experimentalists, dendroclimatologists must, at times, decide to make the best of imperfect data, rather than resample. This tradeoff is made more difficult, because sample collection (in the field) and analysis (in the lab) may be separated significantly in time and space. These collection challenges mean that data gathering is not as simple or cheap as conventional laboratory science. However, they also give the field's practitioners much enjoyment, working out of doors, with hands on trees and tools.

Other Measurements

Initial work focused on measuring the tree ring width—this is simple to measure and can be related to climate parameters. But the annual growth of the tree leaves other traces. In particular *maximum latewood density* (MXD) is another metric used for estimating environmental variables. It is, however, harder to measure. Other properties (e.g. isotope or chemical trace analysis) have also been tried most notably by L. M. Libby in her 1974 paper "Temperature Dependence of Isotope Ratios in Tree Rings". In theory, multiple measurements on the same ring will allow differentiation of confounding factors (e.g. precipitation and temperature). However, most studies are still based on ring widths at limiting stands.

Measuring radiocarbon concentrations in tree rings has proven to be useful in recreating past sunspot activity, with data now extending back over 11,000 years.

References

- Beerling, David (2007). The emerald planet: how plants changed Earth's history. Oxford University press. p. 47. ISBN 9780192806024.

- Chenoweth, Michael (2003). The 18th century climate of Jamaica derived from the journals of Thomas Thistlewood, 1750-1786. Transactions of the American Philosophical Society. 93. Philadelphia: American Philosophical Society. ISBN 0-87169-932-X.

- Grazulis, Thomas P (July 1993). Significant Tornadoes 1680–1991. St. Johnsbury, VT: The Tornado Project of Environmental Films. ISBN 1-879362-03-1.

- Briffa, K.; Cook, E. (1990). "Sect. 5.6: Methods of response function analysis". In Cook, Edward; Kairiūkštis, Leonardas. Methods of Dendrochronology: Applications in the Environmental Sciences. Springer. ISBN 978-0-7923-0586-6.

- Hughes, Malcolm K.; Swetman, Thomas W.; Diaz, Henry, eds. (2010). Dendroclimatology: Progress and Prospects. Springer. ISBN 978-1-4020-4010-8.

- Luckman, B.H. (2007). "Dendroclimatology". In Elias, Scott A. Encyclopedia of Quaternary Science 1. Elsevier. pp. 465–475. ISBN 978-0-444-51919-1.

- Schweingruber, Fritz Hans; Eidgenössische Forschungsanstalt für Wald, Schnee und Landschaft (1996). "Ch. 19". Tree Rings and Environment Dendroecology. Berne: Paul Haupt. ISBN 978-3-258-05458-2.

- "U.S. Tornado Climatology | National Centers for Environmental Information (NCEI) formerly known as National Climatic Data Center (NCDC)". www.ncdc.noaa.gov. Retrieved 2016-04-20.

- "Tornado Alley | National Centers for Environmental Information (NCEI) formerly known as National Climatic Data Center (NCDC)". www.ncdc.noaa.gov. Retrieved 2016-04-20.

- Roth, David M. (April 29, 2015). "Tropical Cyclone Point Maxima". Tropical Cyclone Rainfall Data. United States Weather Prediction Center. Retrieved May 8, 2016.

- Yen, Ben Chie. "Urban Stormwater Modeling and Simulation." Bulletin of the American Meteorological Society Apr. 1995: 564+. Academic OneFile. Web. 11 Nov. 2014.

- Roth, David M; Weather Prediction Center (January 7, 2013). "Maximum Rainfall caused by Tropical Cyclones and their Remnants Per State (1950–2012)". Tropical Cyclone Point Maxima. United States National Oceanic and Atmospheric Administration's National Weather Service. Retrieved March 15, 2013.

- "Page 1 1 International Partnerships in Ice Core Sciences (IPICS) The oldest ice core: A 1.5 million year record of climate and greenhouse gases from Antarctica" (PDF). Retrieved 22 September 2011.

Climate Change: Causes and Effects

Greenhouse gas emissions and the use of aerosols have led to dangerous levels of increase of surface temperature levels on earth. At the same time, climate change can be defined as a great climatic event that influences more than one region. The earth has gone through many climate changes, as has been explained in this chapter.

Climate Change

Climate change is a change in the statistical distribution of weather patterns when that change lasts for an extended period of time (i.e., decades to millions of years). Climate change may refer to a change in average weather conditions, or in the time variation of weather around longer-term average conditions (i.e., more or fewer extreme weather events). Climate change is caused by factors such as biotic processes, variations in solar radiation received by Earth, plate tectonics, and volcanic eruptions. Certain human activities have also been identified as significant causes of recent climate change, often referred to as *global warming*.

Scientists actively work to understand past and future climate by using observations and theoretical models. A climate record—extending deep into the Earth's past—has been assembled, and continues to be built up, based on geological evidence from borehole temperature profiles, cores removed from deep accumulations of ice, floral and faunal records, glacial and periglacial processes, stable-isotope and other analyses of sediment layers, and records of past sea levels. More recent data are provided by the instrumental record. General circulation models, based on the physical sciences, are often used in theoretical approaches to match past climate data, make future projections, and link causes and effects in climate change.

Terminology

The most general definition of *climate change* is a change in the statistical properties (principally its mean and spread) of the climate system when considered over long periods of time, regardless of cause. Accordingly, fluctuations over periods shorter than a few decades, such as El Niño, do not represent climate change.

The term sometimes is used to refer specifically to climate change caused by human activity, as opposed to changes in climate that may have resulted as part of Earth's natural processes. In this sense, especially in the context of environmental policy, the term *climate change* has become synonymous with *anthropogenic global warming*. Within scientific journals, *global warming* refers to surface temperature increases while *climate change* includes global warming and everything else that increasing greenhouse gas levels affect.

Climatic Change Versus Climate Change

In 1966, the World Meteorological Organization (WMO) proposed the term climatic change to encompass all forms of climatic variability on time-scales longer than 10 years, whether the cause was natural or anthropogenic. Change was a given and climatic was used as an adjective to describe this kind of change (as opposed to political or economic change). When it was realized that human activities had a potential to drastically alter the climate, the term climate change replaced climatic change as the dominant term to reflect an anthropogenic cause. Climate change was incorporated in the title of the Intergovernmental Panel on Climate Change (IPCC) and the UN Framework Convention on Climate Change (UNFCCC). Climate change, used as a noun, became an issue rather than the technical description of changing weather.

Causes

On the broadest scale, the rate at which energy is received from the Sun and the rate at which it is lost to space determine the equilibrium temperature and climate of Earth. This energy is distributed around the globe by winds, ocean currents, and other mechanisms to affect the climates of different regions.

Factors that can shape climate are called climate forcings or "forcing mechanisms". These include processes such as variations in solar radiation, variations in the Earth's orbit, variations in the albedo or reflectivity of the continents and oceans, mountain-building and continental drift and changes in greenhouse gas concentrations. There are a variety of climate change feedbacks that can either amplify or diminish the initial forcing. Some parts of the climate system, such as the oceans and ice caps, respond more slowly in reaction to climate forcings, while others respond more quickly. There are also key threshold factors which when exceeded can produce rapid change.

Forcing mechanisms can be either "internal" or "external". Internal forcing mechanisms are natural processes within the climate system itself (e.g., the thermohaline circulation). External forcing mechanisms can be either natural (e.g., changes in solar output) or anthropogenic (e.g., increased emissions of greenhouse gases).

Whether the initial forcing mechanism is internal or external, the response of the climate system might be fast (e.g., a sudden cooling due to airborne volcanic ash reflecting sunlight), slow (e.g. thermal expansion of warming ocean water), or a combination (e.g., sudden loss of albedo in the arctic ocean as sea ice melts, followed by more gradual thermal expansion of the water). Therefore, the climate system can respond abruptly, but the full response to forcing mechanisms might not be fully developed for centuries or even longer.

Internal Forcing Mechanisms

Scientists generally define the five components of earth's climate system to include atmosphere, hydrosphere, cryosphere, lithosphere (restricted to the surface soils, rocks, and sediments), and biosphere. Natural changes in the climate system ("internal forcings") result in internal "climate variability". Examples include the type and distribution of species, and changes in ocean currents.

Ocean Variability

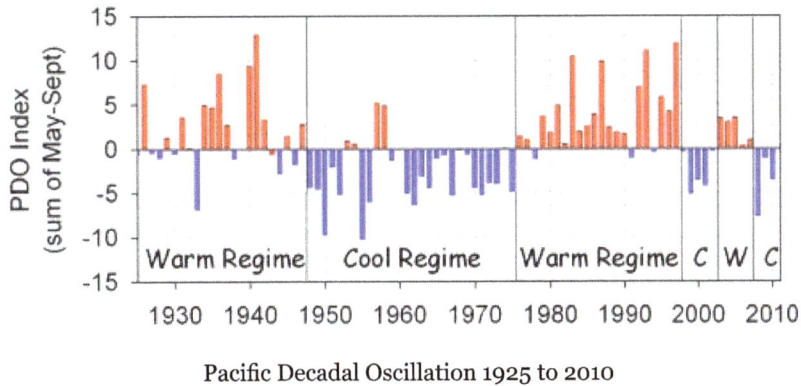

Pacific Decadal Oscillation 1925 to 2010

The ocean is a fundamental part of the climate system, some changes in it occurring at longer timescales than in the atmosphere, as it has hundreds of times more mass and thus very high thermal inertia, with effects such as the ocean depths still lagging today in temperature adjustment from effects of the Little Ice Age of past centuries).

Short-term fluctuations (years to a few decades) such as the El Niño-Southern Oscillation, the Pacific decadal oscillation, the North Atlantic oscillation, and the Arctic oscillation, represent climate variability rather than climate change. On longer time-scales, alterations to ocean processes such as thermohaline circulation play a key role in redistributing heat by carrying out a very slow and extremely deep movement of water and the long-term redistribution of heat in the world's oceans.

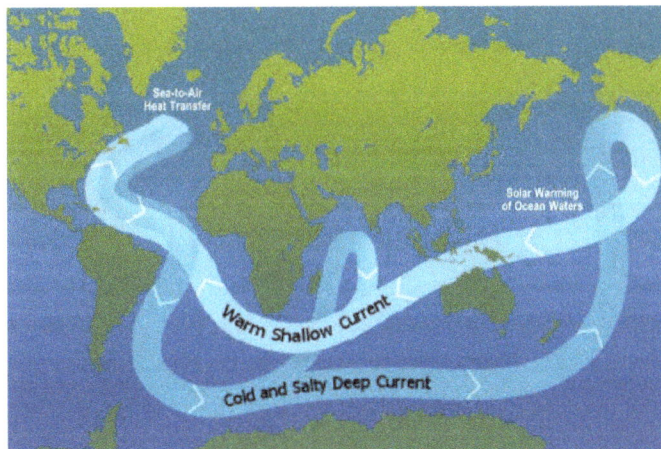

A schematic of modern thermohaline circulation. Tens of millions of years ago, continental-plate movement formed a land-free gap around Antarctica, allowing the formation of the ACC, which keeps warm waters away from Antarctica.

Life

Life affects climate through its role in the carbon and water cycles and through such mechanisms as albedo, evapotranspiration, cloud formation, and weathering. Examples of how life may have affected past climate include:

- glaciation 2.3 billion years ago triggered by the evolution of oxygenic photosynthesis, which depleted the atmosphere of the greenhouse gas carbon dioxide and introduced free oxygen.

- another glaciation 300 million years ago ushered in by long-term burial of decomposition-resistant detritus of vascular land-plants (creating a carbon sink and forming coal)

- termination of the Paleocene-Eocene Thermal Maximum 55 million years ago by flourishing marine phytoplankton

- reversal of global warming 49 million years ago by 800,000 years of arctic azolla blooms

- global cooling over the past 40 million years driven by the expansion of grass-grazer ecosystems

External Forcing Mechanisms

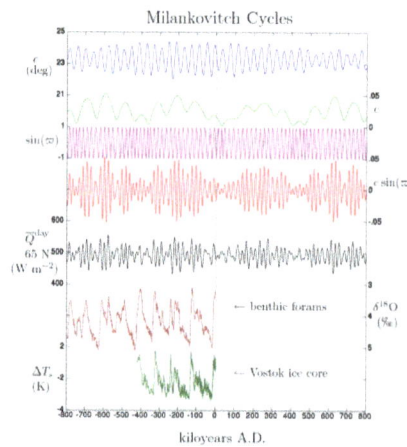

Milankovitch cycles from 800,000 years ago in the past to 800,000 years in the future.

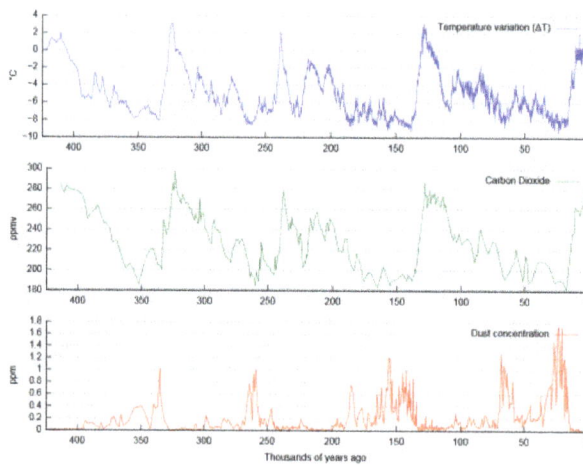

Variations in CO_2, temperature and dust from the Vostok ice core over the last 450,000 years

Orbital Variations

Slight variations in Earth's orbit lead to changes in the seasonal distribution of sunlight reaching the Earth's surface and how it is distributed across the globe. There is very little change to the area-averaged annually averaged sunshine; but there can be strong changes in the geographical and seasonal distribution. The three types of orbital variations are variations in Earth's eccentricity, changes in the tilt angle of Earth's axis of rotation, and precession of Earth's axis. Combined together, these produce Milankovitch cycles which have a large impact on climate and are notable for their correlation to glacial and interglacial periods, their correlation with the advance and retreat of the Sahara, and for their appearance in the stratigraphic record.

The IPCC notes that Milankovitch cycles drove the ice age cycles, CO_2 followed temperature change "with a lag of some hundreds of years," and that as a feedback amplified temperature change. The depths of the ocean have a lag time in changing temperature (thermal inertia on such scale). Upon seawater temperature change, the solubility of CO_2 in the oceans changed, as well as other factors impacting air-sea CO_2 exchange.

Solar Output

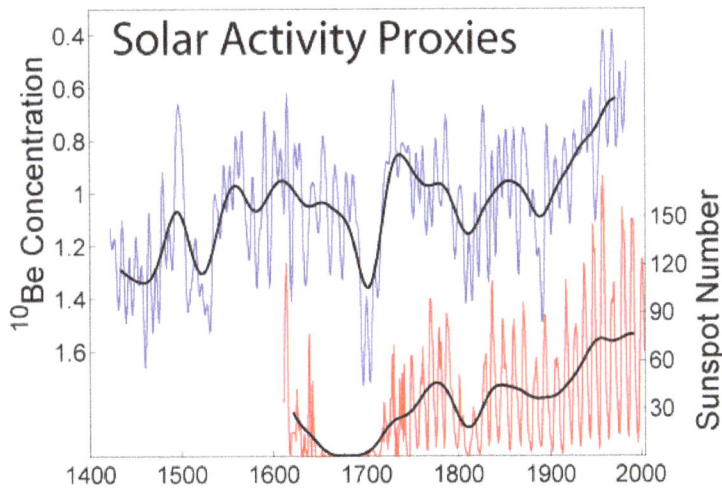

Variations in solar activity during the last several centuries based on observations of sunspots and beryllium isotopes. The period of extraordinarily few sunspots in the late 17th century was the Maunder minimum.

The Sun is the predominant source of energy input to the Earth. Other sources include geothermal energy from the Earth's core, and heat from the decay of radioactive compounds. Both long- and short-term variations in solar intensity are known to affect global climate.

Three to four billion years ago, the Sun emitted only 70% as much power as it does today. If the atmospheric composition had been the same as today, liquid water should not have existed on Earth. However, there is evidence for the presence of water on the early Earth, in the Hadean and Archean eons, leading to what is known as the faint young Sun paradox. Hypothesized solutions to this paradox include a vastly different atmosphere, with much higher concentrations of greenhouse gases than currently exist. Over the following approximately 4 billion years, the energy output of the Sun increased and atmospheric composition changed. The Great Oxygenation Event – oxygenation of the atmosphere around 2.4 billion years ago – was the most notable alteration. Over the next five billion years, the Sun's ultimate death as it becomes a red giant and then a white dwarf will have large effects on climate, with the red giant phase possibly ending any life on Earth that survives until that time.

Solar output also varies on shorter time scales, including the 11-year solar cycle and longer-term modulations. Solar intensity variations possibly as a result of the Wolf, Spörer and Maunder Minimum are considered to have been influential in triggering the Little Ice Age, and some of the warming observed from 1900 to 1950. The cyclical nature of the Sun's energy output is not yet fully understood; it differs from the very slow change that is happening within the Sun as it ages and evolves. Research indicates that solar variability has had effects including the Maunder minimum from 1645 to 1715 A.D., part of the Little Ice Age from 1550 to 1850 A.D. that was marked by relative cooling and greater glacier extent than the centuries before and afterward. Some studies point toward solar radiation increases from cyclical sunspot activity affecting global warming, and climate may be influenced by the sum of all effects (solar variation, anthropogenic radiative forcings, etc.).

Interestingly, a 2010 study *suggests*, "that the effects of solar variability on temperature throughout the atmosphere may be contrary to current expectations."

In an Aug 2011 Press Release, CERN announced the publication in the Nature journal the initial results from its CLOUD experiment. The results indicate that ionisation from cosmic rays significantly enhances aerosol formation in the presence of sulfuric acid and water, but in the lower atmosphere where ammonia is also required, this is insufficient to account for aerosol formation and additional trace vapours must be involved. The next step is to find more about these trace vapours, including whether they are of natural or human origin.

Volcanism

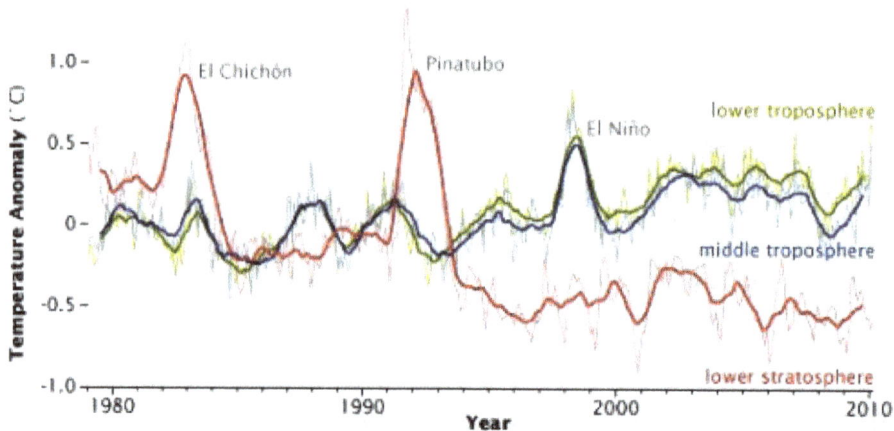

In atmospheric temperature from 1979 to 2010, determined by MSU NASA satellites, effects appear from aerosols released by major volcanic eruptions (El Chichón and Pinatubo). El Niño is a separate event, from ocean variability.

The eruptions considered to be large enough to affect the Earth's climate on a scale of more than 1 year are the ones that inject over 100,000 tons of SO_2 into the stratosphere. This is due to the optical properties of SO_2 and sulfate aerosols, which strongly absorb or scatter solar radiation, creating a global layer of sulfuric acid haze. On average, such eruptions occur several times per century, and cause cooling (by partially blocking the transmission of solar radiation to the Earth's surface) for a period of a few years.

The eruption of Mount Pinatubo in 1991, the second largest terrestrial eruption of the 20th century, affected the climate substantially, subsequently global temperatures decreased by about 0.5 °C (0.9 °F) for up to three years. Thus, the cooling over large parts of the Earth reduced surface temperatures in 1991-93, the equivalent to a reduction in net radiation of 4 watts per square meter.

The Mount Tambora eruption in 1815 caused the Year Without a Summer. Much larger eruptions, known as large igneous provinces, occur only a few times every fifty - hundred million years - through flood basalt, and caused in Earth past global warming and mass extinctions.

Small eruptions, with injections of less than 0.1 Mt of sulfur dioxide into the stratosphere, impact the atmosphere only subtly, as temperature changes are comparable with natural variability. However, because smaller eruptions occur at a much higher frequency, they too have a significant impact on Earth's atmosphere.

Seismic monitoring maps current and future trends in volcanic activities, and tries to develop early warning systems. In climate modelling the aim is to study the physical mechanisms and feedbacks of volcanic forcing.

Volcanoes are also part of the extended carbon cycle. Over very long (geological) time periods, they release carbon dioxide from the Earth's crust and mantle, counteracting the uptake by sedimentary rocks and other geological carbon dioxide sinks. The US Geological Survey estimates are that volcanic emissions are at a much lower level than the effects of current human activities, which generate 100–300 times the amount of carbon dioxide emitted by volcanoes. A review of published studies indicates that annual volcanic emissions of carbon dioxide, including amounts released from mid-ocean ridges, volcanic arcs, and hot spot volcanoes, are only the equivalent of 3 to 5 days of human caused output. The annual amount put out by human activities may be greater than the amount released by supererruptions, the most recent of which was the Toba eruption in Indonesia 74,000 years ago.

Although volcanoes are technically part of the lithosphere, which itself is part of the climate system, the IPCC explicitly defines volcanism as an external forcing agent.

Plate Tectonics

Over the course of millions of years, the motion of tectonic plates reconfigures global land and ocean areas and generates topography. This can affect both global and local patterns of climate and atmosphere-ocean circulation.

The position of the continents determines the geometry of the oceans and therefore influences patterns of ocean circulation. The locations of the seas are important in controlling the transfer of heat and moisture across the globe, and therefore, in determining global climate. A recent example of tectonic control on ocean circulation is the formation of the Isthmus of Panama about 5 million years ago, which shut off direct mixing between the Atlantic and Pacific Oceans. This strongly affected the ocean dynamics of what is now the Gulf Stream and may have led to Northern Hemisphere ice cover. During the Carboniferous period, about 300 to 360 million years ago, plate tectonics may have triggered large-scale storage of carbon and increased glaciation. Geologic evidence points to a "megamonsoonal" circulation pattern during the time of the supercontinent Pangaea, and climate modeling suggests that the existence of the supercontinent was conducive to the establishment of monsoons.

The size of continents is also important. Because of the stabilizing effect of the oceans on temperature, yearly temperature variations are generally lower in coastal areas than they are inland. A larger supercontinent will therefore have more area in which climate is strongly seasonal than will several smaller continents or islands.

Human Influences

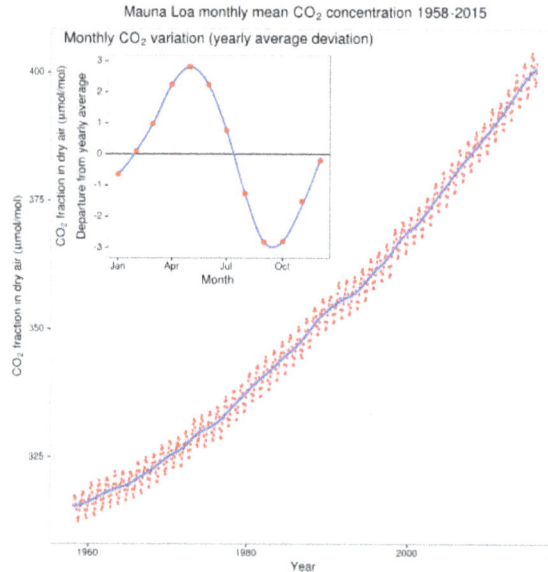

Increase in atmospheric CO_2 levels

In the context of climate variation, anthropogenic factors are human activities which affect the climate. The scientific consensus on climate change is "that climate is changing and that these changes are in large part caused by human activities," and it "is largely irreversible."

"Science has made enormous inroads in understanding climate change and its causes, and is beginning to help develop a strong understanding of current and potential impacts that will affect people today and in coming decades. This understanding is crucial because it allows decision makers to place climate change in the context of other large challenges facing the nation and the world. There are still some uncertainties, and there always will be in understanding a complex system like Earth's climate. Nevertheless, there is a strong, credible body of evidence, based on multiple lines of research, documenting that climate is changing and that these changes are in large part caused by human activities. While much remains to be learned, the core phenomenon, scientific questions, and hypotheses have been examined thoroughly and have stood firm in the face of serious scientific debate and careful evaluation of alternative explanations."

— United States National Research Council, Advancing the Science of Climate Change

Of most concern in these anthropogenic factors is the increase in CO_2 levels due to emissions from fossil fuel combustion, followed by aerosols (particulate matter in the atmosphere) and the CO_2 released by cement manufacture. Other factors, including land use, ozone depletion, animal agriculture and deforestation, are also of concern in the roles they play – both separately and in conjunction with other factors – in affecting climate, microclimate, and measures of climate variables.

Physical Evidence

Evidence for climatic change is taken from a variety of sources that can be used to reconstruct past climates. Reasonably complete global records of surface temperature are available beginning from the mid-late 19th century. For earlier periods, most of the evidence is indirect—climatic changes are inferred from changes in proxies, indicators that reflect climate, such as vegetation, ice cores, dendrochronology, sea level change, and glacial geology.

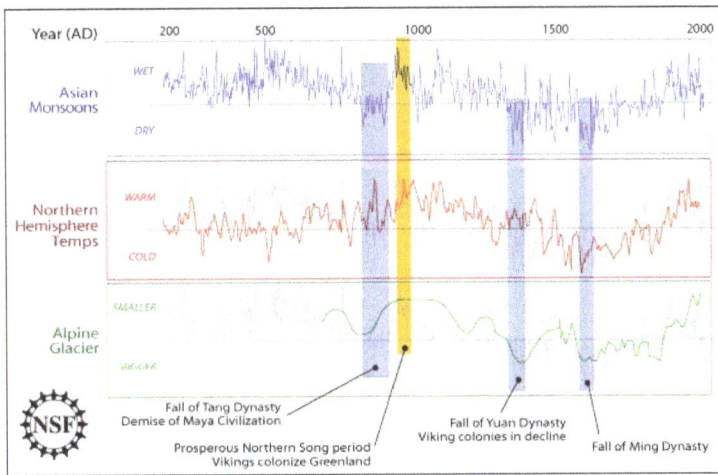

Comparisons between Asian Monsoons from 200 AD to 2000 AD (staying in the background on other plots), Northern Hemisphere temperature, Alpine glacier extent (vertically inverted as marked), and human history as noted by the U.S. NSF.

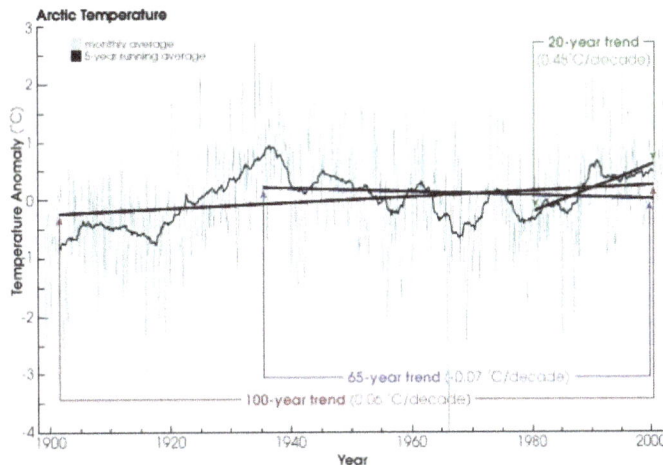

Arctic temperature anomalies over a 100-year period as estimated by NASA. Typical high monthly variance can be seen, while longer-term averages highlight trends.

Temperature Measurements and Proxies

The instrumental temperature record from surface stations was supplemented by radiosonde balloons, extensive atmospheric monitoring by the mid-20th century, and, from the 1970s on, with global satellite data as well. The $^{18}O/^{16}O$ ratio in calcite and ice core samples used to deduce ocean temperature in the distant past is an example of a temperature proxy method, as are other climate metrics noted in subsequent categories.

Historical and Archaeological Evidence

Climate change in the recent past may be detected by corresponding changes in settlement and agricultural patterns. Archaeological evidence, oral history and historical documents can offer insights into past changes in the climate. Climate change effects have been linked to the collapse of various civilizations.

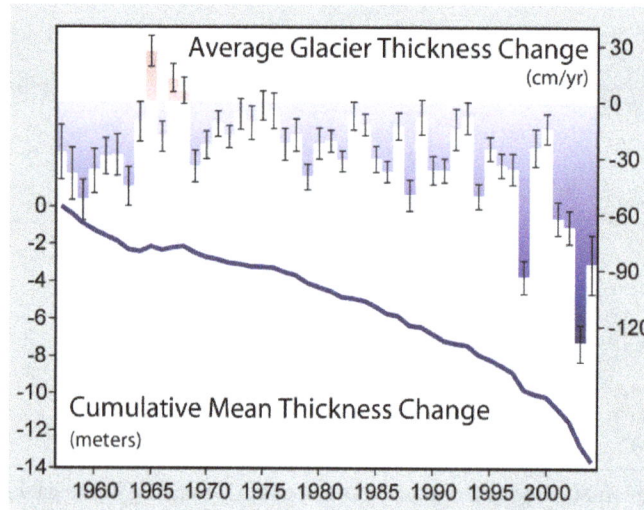

Decline in thickness of glaciers worldwide over the past half-century

Glaciers

Glaciers are considered among the most sensitive indicators of climate change. Their size is determined by a mass balance between snow input and melt output. As temperatures warm, glaciers retreat unless snow precipitation increases to make up for the additional melt; the converse is also true.

Glaciers grow and shrink due both to natural variability and external forcings. Variability in temperature, precipitation, and englacial and subglacial hydrology can strongly determine the evolution of a glacier in a particular season. Therefore, one must average over a decadal or longer time-scale and/or over many individual glaciers to smooth out the local short-term variability and obtain a glacier history that is related to climate.

A world glacier inventory has been compiled since the 1970s, initially based mainly on aerial photographs and maps but now relying more on satellites. This compilation tracks more than 100,000 glaciers covering a total area of approximately 240,000 km², and preliminary estimates indicate that the remaining ice cover is around 445,000 km². The World Glacier Monitoring Service collects data annually on glacier retreat and glacier mass balance. From this data, glaciers worldwide have been found to be shrinking significantly, with strong glacier retreats in the 1940s, stable or growing conditions during the 1920s and 1970s, and again retreating from the mid-1980s to present.

The most significant climate processes since the middle to late Pliocene (approximately 3 million years ago) are the glacial and interglacial cycles. The present interglacial period (the Holocene) has lasted about 11,700 years. Shaped by orbital variations, responses such as the rise and fall of continental ice sheets and significant sea-level changes helped create the climate. Other changes, including Heinrich events, Dansgaard–Oeschger events and the Younger Dryas, however, illustrate how glacial variations may also influence climate without the orbital forcing.

Glaciers leave behind moraines that contain a wealth of material—including organic matter, quartz, and potassium that may be dated—recording the periods in which a glacier advanced and retreated. Similarly, by tephrochronological techniques, the lack of glacier cover can be identified by the presence of soil or volcanic tephra horizons whose date of deposit may also be ascertained.

Arctic Sea Ice Loss

The decline in Arctic sea ice, both in extent and thickness, over the last several decades is further evidence for rapid climate change. Sea ice is frozen seawater that floats on the ocean surface. It covers millions of square miles in the polar regions, varying with the seasons. In the Arctic, some sea ice remains year after year, whereas almost all Southern Ocean or Antarctic sea ice melts away and reforms annually. Satellite observations show that Arctic sea ice is now declining at a rate of 13.3 percent per decade, relative to the 1981 to 2010 average.

Vegetation

A change in the type, distribution and coverage of vegetation may occur given a change in the climate. Some changes in climate may result in increased precipitation and warmth, resulting in improved plant growth and the subsequent sequestration of airborne CO_2. A gradual increase in warmth in a region will lead to earlier flowering and fruiting times, driving a change in the timing of life cycles of dependent organisms. Conversely, cold will cause plant bio-cycles to lag. Larger, faster or more radical changes, however, may result in vegetation stress, rapid plant loss and desertification in certain circumstances. An example of this occurred during the Carboniferous Rainforest Collapse (CRC), an extinction event 300 million years ago. At this time vast rainforests covered the equatorial region of Europe and America. Climate change devastated these tropical rainforests, abruptly fragmenting the habitat into isolated 'islands' and causing the extinction of many plant and animal species.

Pollen Analysis

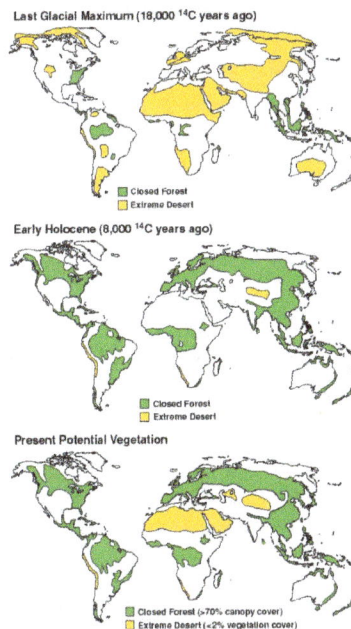

Top: Arid ice age climate
Middle: Atlantic Period, warm and wet
Bottom: Potential vegetation in climate now if not for human effects like agriculture.

Palynology is the study of contemporary and fossil palynomorphs, including pollen. Palynology is used to infer the geographical distribution of plant species, which vary under different climate conditions. Different groups of plants have pollen with distinctive shapes and surface textures, and since the outer surface of pollen is composed of a very resilient material, they resist decay. Changes in the type of pollen found in different layers of sediment in lakes, bogs, or river deltas indicate changes in plant communities. These changes are often a sign of a changing climate. As an example, palynological studies have been used to track changing vegetation patterns throughout the Quaternary glaciations and especially since the last glacial maximum.

Cloud Cover and Precipitation

Past precipitation can be estimated in the modern era with the global network of precipitation gauges. Surface coverage over oceans and remote areas is relatively sparse, but, reducing reliance on interpolation, satellite clouds and precipitation data has been available since the 1970s. Quantification of climatological variation of precipitation in prior centuries and epochs is less complete but approximated using proxies such as marine sediments, ice cores, cave stalagmites, and tree rings. In July 2016 scientists published evidence of increased cloud cover over polar regions, as predicted by climate models.

Climatological temperatures substantially affect cloud cover and precipitation. For instance, during the Last Glacial Maximum of 18,000 years ago, thermal-driven evaporation from the oceans onto continental landmasses was low, causing large areas of extreme desert, including polar deserts (cold but with low rates of cloud cover and precipitation). In contrast, the world's climate was cloudier and wetter than today near the start of the warm Atlantic Period of 8000 years ago.

Estimated global land precipitation increased by approximately 2% over the course of the 20th century, though the calculated trend varies if different time endpoints are chosen, complicated by ENSO and other oscillations, including greater global land cloud cover precipitation in the 1950s and 1970s than the later 1980s and 1990s despite the positive trend over the century overall. Similar slight overall increase in global river runoff and in average soil moisture has been perceived.

Dendroclimatology

Dendroclimatology is the analysis of tree ring growth patterns to determine past climate variations. Wide and thick rings indicate a fertile, well-watered growing period, whilst thin, narrow rings indicate a time of lower rainfall and less-than-ideal growing conditions.

Ice Cores

Analysis of ice in a core drilled from an ice sheet such as the Antarctic ice sheet, can be used to show a link between temperature and global sea level variations. The air trapped in bubbles in the ice can also reveal the CO_2 variations of the atmosphere from the distant past, well before modern environmental influences. The study of these ice cores has been a significant indicator of the changes in CO_2 over many millennia, and continues to provide valuable information about the differences between ancient and modern atmospheric conditions.

Animals

Remains of beetles are common in freshwater and land sediments. Different species of beetles tend to be found under different climatic conditions. Given the extensive lineage of beetles whose genetic makeup has not altered significantly over the millennia, knowledge of the present climatic range of the different species, and the age of the sediments in which remains are found, past climatic conditions may be inferred.

Similarly, the historical abundance of various fish species has been found to have a substantial relationship with observed climatic conditions. Changes in the primary productivity of autotrophs in the oceans can affect marine food webs.

Sea Level Change

Global sea level change for much of the last century has generally been estimated using tide gauge measurements collated over long periods of time to give a long-term average. More recently, altimeter measurements — in combination with accurately determined satellite orbits — have provided an improved measurement of global sea level change. To measure sea levels prior to instrumental measurements, scientists have dated coral reefs that grow near the surface of the ocean, coastal sediments, marine terraces, ooids in limestones, and nearshore archaeological remains. The predominant dating methods used are uranium series and radiocarbon, with cosmogenic radionuclides being sometimes used to date terraces that have experienced relative sea level fall. In the early Pliocene, global temperatures were 1–2°C warmer than the present temperature, yet sea level was 15–25 meters higher than today.

Climate of Recent Glaciations

- Bond event

Climate of the Past

- Ice ages

- Paleocene–Eocene Thermal Maximum

- Permo-Carboniferous Glaciation

- Snowball Earth

Recent Climate

- Anthropocene

- CORA dataset temperature and salinity of global oceans

- Effects of global warming on oceans

- Extreme weather

- Land surface effects on climate

- Hardiness zone

- Holocene climatic optimum

- Medieval Warm Period

- Temperature record of the past 1000 years

Climate Engineering

Climate engineering, commonly referred to as geoengineering, also known as climate intervention, is the deliberate and large-scale intervention in the Earth's climatic system with the aim of limiting adverse climate change. Climate engineering is an umbrella term for two types of measures: carbon dioxide removal and solar radiation management. Carbon dioxide removal addresses the cause of climate change by removing one of the greenhouse gases (carbon dioxide) from the atmosphere. Solar radiation management attempts to offset effects of greenhouse gases by causing the Earth to absorb less solar radiation.

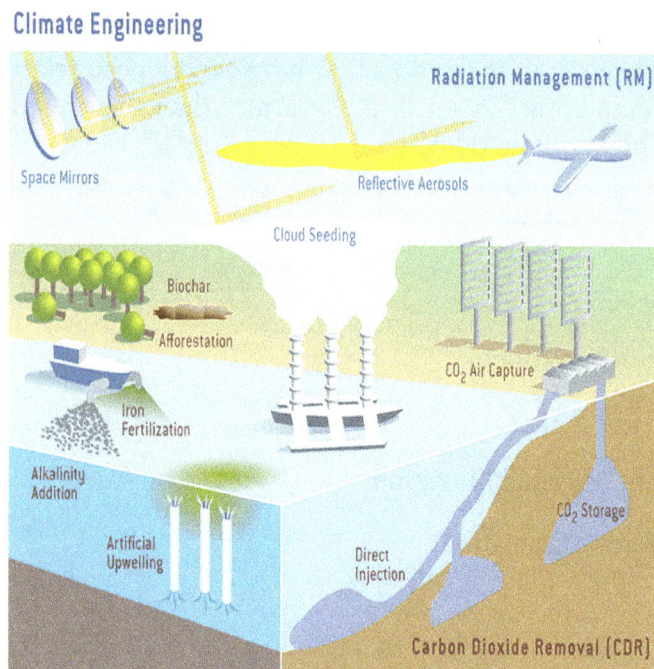

Diagram of several proposed climate engineering methods.

Climate engineering approaches are sometimes viewed as additional potential options for limiting climate change, alongside mitigation and adaptation. There is substantial agreement among scientists that climate engineering cannot substitute for climate change mitigation. Some approaches might be used as accompanying measures to sharp cuts in greenhouse gas emissions. Given that all types of measures for addressing climate change have economic, political, or physical limitations as some climate engineering approaches might eventually be used as part of an ensemble of measures. Research on costs, benefits, and various types of risks of most climate engineering approaches is at an early stage and their understanding needs to improve to judge their adequacy and feasibility.

No outdoor solar radiation management projects have taken place to date. Almost all research into solar radiation management has consisted of computer modelling or laboratory tests, and an attempt to move to outdoor experimentation was controversial. Some carbon dioxide removal practices, such as planting of trees and bio-energy with carbon capture and storage projects, are underway. Their scalability to effectively affect global climate is however debated. Ocean iron fertilization has been given small-scale research trials, sparking substantial controversy.

Most experts and major reports advise against relying on climate engineering techniques as a simple solution to climate change, in part due to the large uncertainties over effectiveness and side effects. However, most experts also argue that the risks of such interventions must be seen in the context of risks of dangerous climate change. Interventions at large scale may run a greater risk of disrupting natural systems resulting in a dilemma that those approaches that could prove highly (cost-) effective in addressing extreme climate risk, might themselves cause substantial risk. Some have suggested that the concept of engineering the climate presents a so-called "moral hazard" because it could reduce political and public pressure for emissions reduction, which could exacerbate overall climate risks; others assert that the threat of climate engineering could spur emissions cuts. Groups such as ETC Group and some climate researchers (such as Raymond Pierrehumbert) are in favour of a moratorium on out-of-doors testing and deployment of solar radiation management (SRM.)

Background

With respect to climate, geoengineering is defined by the Royal Society as "... the deliberate large-scale intervention in the Earth's climate system, in order to moderate global warming."

Several organizations have investigated climate engineering with a view to evaluating its potential, including the US Congress, the National Academy of Sciences, the Royal Society, and the UK Parliament. The Asilomar International Conference on Climate Intervention Technologies was convened to identify and develop risk reduction guidelines for climate intervention experimentation.

Some environmental organisations (such as Friends of the Earth and Greenpeace) have been reluctant to endorse solar radiation management, but are often more supportive of some carbon dioxide removal projects, such as afforestation and peatland restoration. Some authors have argued that any public support for climate engineering may weaken the fragile political consensus to reduce greenhouse gas emissions.

Proposed Strategies

Several climate engineering strategies have been proposed. IPCC documents detail several notable proposals. These fall into two main categories: solar radiation management and carbon dioxide removal. Here is a list of specific proposals.

Solar Radiation Management

Solar radiation management (SRM) techniques would seek to reduce sunlight absorbed (ultra-violet, near infra-red and visible). This would be achieved by deflecting sunlight away from the Earth, or by increasing the reflectivity (albedo) of the atmosphere or the Earth's surface. These methods would not reduce greenhouse gas concentrations in the atmosphere, and thus would not seek to address problems such as the ocean acidification caused by CO_2. In general, solar radiation management projects presently appear to be able to take effect rapidly and to have very low direct implementation costs relative to greenhouse gas emissions cuts and carbon dioxide removal. Furthermore, many proposed SRM methods would be reversible in their direct climatic effects. While greenhouse gas remediation offers a more comprehensive possible solution to climate change, it does not give instantaneous results; for that, solar radiation management is required.

Solar radiation management methods may include:

- Surface-based: for example, using pale-colored roofing materials, attempting to change the oceans' brightness, or growing high-albedo crops.

- Troposphere-based: for example, marine cloud brightening, which would spray fine sea water to whiten clouds and thus increase cloud reflectivity.

- Upper atmosphere-based: creating reflective aerosols, such as stratospheric sulfate aerosols, specifically designed self-levitating aerosols, or other substances.

- Space-based: space sunshade—obstructing solar radiation with space-based mirrors, dust, etc.

Carbon Dioxide Removal

An oceanic phytoplankton bloom in the South Atlantic Ocean, off the coast of Argentina. The aim of ocean iron fertilization in theory is to increase such blooms by adding some iron, which would then draw carbon from the atmosphere and fix it on the seabed.

Carbon dioxide removal (sometimes known as negative emissions technologies or greenhouse gas removal) projects seek to remove carbon dioxide from the atmosphere. Proposed methods include those that directly remove such gases from the atmosphere, as well as indirect methods that seek to promote natural processes that draw down and sequester CO_2 (e.g. tree planting). Many projects overlap with carbon capture and storage projects, and may not be considered to be climate engineering by all commentators. Techniques in this category include:

- Creating biochar, which can be mixed with soil to create terra preta

- Bio-energy with carbon capture and storage to sequester carbon and simultaneously provide energy

- Carbon air capture to remove carbon dioxide from ambient air

- Afforestation, reforestation and forest restoration to absorb carbon dioxide

- Ocean fertilization including iron fertilisation of the oceans

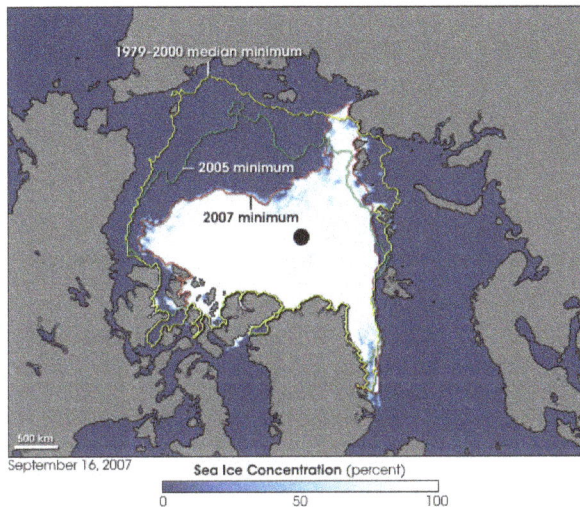

Significant reduction in ice volume in the Arctic Ocean in the range between 1979 and 2007 years

Justification

Tipping Points and Positive Feedback

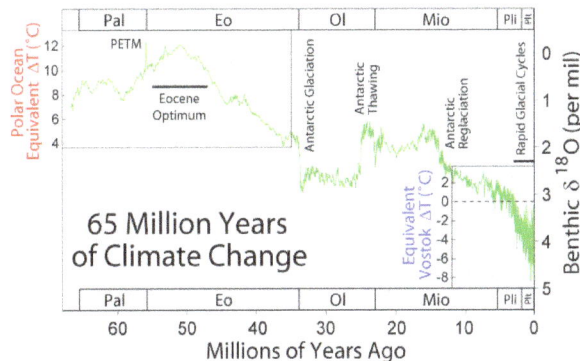

Climate change during the last 65 million years. The Paleocene–Eocene Thermal Maximum is labelled PETM.

It is argued that climate change may cross tipping points where elements of the climate system may 'tip' from one stable state to another stable state, much like a glass tipping over. When the new state is reached, further warming may be caused by positive feedback effects,. An example of a proposed causal chain leading to runaway global warming is the collapse of Arctic sea ice triggering subsequent release of methane. Such a scenario, however, is regarded as unlikely by many scientists.

The precise identity of such "tipping points" is not clear, with scientists taking differing views on whether specific systems are capable of "tipping" and the point at which this "tipping" will occur. An example of a previous tipping point is that which preceded the rapid warming leading up to the Paleocene–Eocene Thermal Maximum. Once a tipping point is crossed, cuts in anthropogenic greenhouse gas emissions will not be able to reverse the change. Conservation of resources and reduction of greenhouse emissions, used in conjunction with climate engineering, are therefore considered a viable option by some commentators.

Buying Time

Climate engineering offers the hope of temporarily reversing some aspects of climate change and allowing the natural climate to be substantially preserved whilst greenhouse gas emissions are brought under control and removed from the atmosphere by natural or artificial processes.

Costs

Estimates of direct costs for climate engineering implementation vary widely. In general, carbon dioxide removal methods are more expensive than the solar radiation management ones. In their 2009 report *Geoengineering the Climate* the Royal Society judged afforestation and stratospheric aerosol injection as the methods with the "highest affordability" (lowest costs). More recently, research into costs of solar radiation management have been published. This suggests that "well designed systems" might be available for costs in the order of a few hundred million to tens of billions of dollars per year. These are much lower than costs to achieve comprehensive reductions in CO_2 emissions. Such costs would be within the budget of most nations, and even some wealthy individuals.

Ethics and Responsibility

Climate engineering would represent a large-scale, intentional effort to modify the climate. It would differ from activities such as burning fossil fuels, as they change the climate inadvertently. Intentional climate change is often viewed differently from a moral standpoint. It raises questions of whether humans have the right to change the climate deliberately, and under what conditions. For example, there may be an ethical distinction between climate engineering to minimize anthropogenic climate change and doing so to optimize the climate. Furthermore, ethical arguments often confront larger considerations of worldview, including individual and social religious commitments. This may imply that discussions of climate engineering should reflect on how religious commitments might influence the discourse. For many people, religious beliefs are pivotal in defining the role of human beings in the wider world. Some religious communities might claim that humans have no responsibility in managing the climate, instead seeing such world systems as the exclusive domain of a Creator. In contrast, other religious communities might see the human role as one of "stewardship" or benevolent management of the world. The question of ethics also relates to issues of policy decision-making. For example, the selection of a globally agreed target temperature is a significant problem in any climate engineering governance regime, as different countries or interest groups may seek different global temperatures.

Politics

It has been argued that regardless of the economic, scientific and technical aspects, the difficulty of achieving concerted political action on climate change requires other approaches. Those arguing political expediency say the difficulty of achieving meaningful emissions cuts and the effective failure of the Kyoto Protocol demonstrate the practical difficulties of achieving carbon dioxide emissions reduction by the agreement of the international community. However, others point to support for climate engineering proposals among think tanks with a history of climate change skepticism and opposition to emissions reductions as evidence that the prospect of climate engineering is itself already politicized and being promoted as part of an argument against the need for (and viability of) emissions reductions; that, rather than climate engineering being a solution to the difficulties of emissions reductions, the prospect of climate engineering is being used as part of an argument to stall emissions reductions in the first place.

Risks and Criticisms

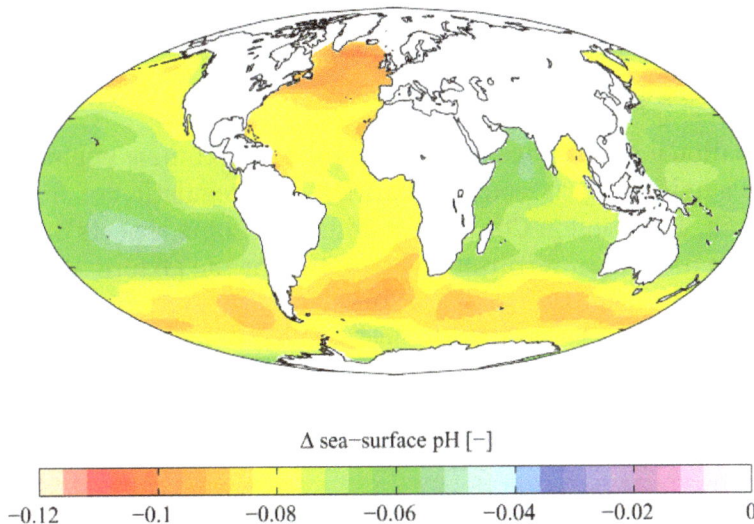

Change in sea surface pH caused by anthropogenic CO_2 between the 1700s and the 1990s. This ocean acidification will still be a major problem unless atmospheric CO_2 is reduced.

Various criticisms have been made of climate engineering, particularly solar radiation management (SRM) methods. Decision making suffers from intransitivity of policy choice. Some commentators appear fundamentally opposed. Groups such as ETC Group and individuals such as Raymond Pierrehumbert have called for a moratorium on climate engineering techniques.

Ineffectiveness

The effectiveness of the techniques proposed may fall short of predictions. In ocean iron fertilization, for example, the amount of carbon dioxide removed from the atmosphere may be much lower than predicted, as carbon taken up by plankton may be released back into the atmosphere from dead plankton, rather than being carried to the bottom of the sea and sequestered.

Moral Hazard or Risk Compensation

The existence of such techniques may reduce the political and social impetus to reduce carbon emissions. This has generally been called a potential moral hazard, although risk compensation may be a more accurate term. This concern causes many environmental groups and campaigners to be reluctant to advocate or discuss climate engineering for fear of reducing the imperative to cut greenhouse gas emissions. However, several public opinion surveys and focus groups have found evidence of either assertions of a desire to increase emission cuts in the face of climate engineering, or of no effect. Other modelling work suggests that the threat of climate engineering may in fact increase the likelihood of emissions reduction.

Governance

Climate engineering opens up various political and economic issues. The governance issues characterizing carbon dioxide removal compared to solar radiation management tend to be distinct. Carbon dioxide removal techniques are typically slow to act, expensive, and entail risks that are relatively familiar, such as the risk of carbon dioxide leakage from underground storage formations. In contrast, solar radiation management methods are fast-acting, comparatively cheap, and involve novel and more significant risks such as regional climate disruptions. As a result of these differing characteristics, the key governance problem for carbon dioxide removal (as with emissions reductions) is making sure actors do enough of it (the so-called "free rider problem"), whereas the key governance issue for solar radiation management is making sure actors do not do too much (the "free driver" problem).

Domestic and international governance vary by the proposed climate engineering method. There is presently a lack of a universally agreed framework for the regulation of either climate engineering activity or research. The London Convention addresses some aspects of the law in relation to biomass ocean storage and ocean fertilization. Scientists at the Oxford Martin School at Oxford University have proposed a set of voluntary principles, which may guide climate engineering research. The short version of the 'Oxford Principles' is:

- Principle 1: Geoengineering to be regulated as a public good.

- Principle 2: Public participation in geoengineering decision-making

- Principle 3: Disclosure of geoengineering research and open publication of results

- Principle 4: Independent assessment of impacts

- Principle 5: Governance before deployment

These principles have been endorsed by the House of Commons of the United Kingdom Science and Technology Select Committee on "The Regulation of Geoengineering", and have been referred to by authors discussing the issue of governance.

The Asilomar conference was replicated to deal with the issue of climate engineering governance, and covered in a TV documentary, broadcast in Canada.

Implementation Issues

There is general consensus that no climate engineering technique is currently sufficiently safe or effective to greatly reduce climate change risks, for the reasons listed above. However, some may be able to contribute to reducing climate risks within relatively short times.

All proposed solar radiation management techniques require implementation on a relatively large scale, in order to impact the Earth's climate. The least costly proposals are budgeted at tens of billions of US dollars annually. Space sunshades would cost far more. Who was to bear the substantial costs of some climate engineering techniques may be hard to agree. However, the more effective solar radiation management proposals currently appear to have low enough direct implementation costs that it would be in the interests of several single countries to implement them unilaterally.

In contrast, carbon dioxide removal, like greenhouse gas emissions reductions, have impacts proportional to their scale. These techniques would not be "implemented" in the same sense as solar radiation management ones.The problem structure of carbon dioxide removal resembles that of emissions cuts, in that both are somewhat expensive public goods, whose provision presents a collective action problem.

Before they are ready to be used, most techniques would require technical development processes that are not yet in place. As a result, many promising proposed climate engineering do not yet have the engineering development or experimental evidence to determine their feasibility or efficacy.

Evaluation of Climate Engineering

Most of what is known about the suggested techniques is based on laboratory experiments, observations of natural phenomena, and on computer modelling techniques. Some proposed climate engineering methods employ methods that have analogues in natural phenomena such as stratospheric sulfur aerosols and cloud condensation nuclei. As such, studies about the efficacy of these methods can draw on information already available from other research, such as that following the 1991 eruption of Mount Pinatubo. However, comparative evaluation of the relative merits of each technology is complicated, especially given modelling uncertainties and the early stage of engineering development of many proposed climate engineering methods .

Reports into climate engineering have also been published in the United Kingdom by the Institution of Mechanical Engineers and the Royal Society. The IMechE report examined a small subset of proposed methods (air capture, urban albedo and algal-based CO_2 capture techniques), and its main conclusions were that climate engineering should be researched and trialled at the small scale alongside a wider decarbonisation of the economy.

The Royal Society review examined a wide range of proposed climate engineering methods and evaluated them in terms of effectiveness, affordability, timeliness and safety (assigning qualitative estimates in each assessment). The report divided proposed methods into "carbon dioxide removal" (CDR) and "solar radiation management" (SRM) approaches that respectively address longwave and shortwave radiation. The key recommendations of the report were that "Parties to the UNFCCC should make increased efforts towards mitigating and adapting to climate change, and in particular to agreeing to global emissions reductions", and that "[nothing] now known about climate engineering options gives any reason to diminish these efforts". Nonetheless, the report also recommended that "research and development of climate engineering options should be undertaken to investigate whether low risk methods can be made available if it becomes necessary to reduce the rate of warming this century".

In a 2009 review study, Lenton and Vaughan evaluated a range of proposed climate engineering techniques from those that sequester CO_2 from the atmosphere and decrease longwave radiation trapping, to those that decrease the Earth's receipt of shortwave radiation. In order to permit a comparison of disparate techniques, they used a common evaluation for each technique based on its effect on net radiative forcing. As such, the review examined the scientific plausibility of proposed methods rather than the practical considerations such as engineering feasibility or economic cost. Lenton and Vaughan found that "[air] capture and storage shows the greatest potential, combined with afforestation, reforestation and bio-char production", and noted that "other suggestions that have received considerable media attention, in particular "ocean pipes" appear to be ineffective". They concluded that "[climate] geoengineering is best considered as a potential complement to the mitigation of CO_2 emissions, rather than as an alternative to it".

In October 2011, a Bipartisan Policy Center panel issued a report urging immediate researching and testing in case "the climate system reaches a 'tipping point' and swift remedial action is required".

National Academy of Sciences

The National Academy of Sciences conducted a 21-month project to study the potential impacts, benefits, and costs of two different types of climate engineering: carbon dioxide removal and albedo modification (solar radiation management). The differences between these two classes of climate engineering "led the committee to evaluate the two types of approaches separately in companion reports, a distinction it hopes carries over to future scientific and policy discussions."

According to the two-volume study released in February 2015:

"Climate intervention is no substitute for reductions in carbon dioxide emissions and adaptation efforts aimed at reducing the negative consequences of climate change. However, as our planet enters a period of changing climate never before experienced in recorded human history, interest is growing in the potential for deliberate intervention in the climate system to counter climate change. ...Carbon dioxide removal strategies address a key driver of climate change, but research is needed to fully assess if any of these technologies could be appropriate for large-scale deployment. Albedo modification strategies could rapidly cool the planet's surface but pose environmental and other risks that are not well understood and therefore should not be deployed at climate-altering scales; more research is needed to determine if albedo modification approaches could be viable in the future."

The project was sponsored by the National Academy of Sciences, U.S. Intelligence Community, National Oceanic and Atmospheric Administration, NASA, and U.S. Department of Energy.

Intergovernmental Panel On Climate Change

The Intergovernmental Panel on Climate Change (IPCC) assessed the scientific literature on climate engineering (referred to as "geoengineering" in its reports), in which it considered carbon dioxide removal and solar radiation separately. Its Fifth Assessment Report states:

Models consistently suggest that SRM would generally reduce climate differences compared to a world with elevated GHG concentrations and no SRM; however, there would also be residual regional differences in climate (e.g., temperature and rainfall) when compared to a climate without elevated GHGs....

Models suggest that if SRM methods were realizable they would be effective in countering increasing temperatures, and would be less, but still, effective in countering some other climate changes. SRM would not counter all effects of climate change, and all proposed geoengineering methods also carry risks and side effects. Additional consequences cannot yet be anticipated as the level of scientific understanding about both SRM and CDR is low. There are also many (political, ethical, and practical) issues involving geoengineering that are beyond the scope of this report.

Climate Sensitivity

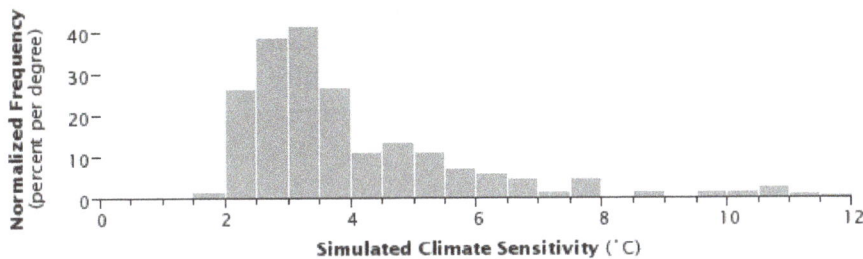

Frequency distribution of climate sensitivity, based on model simulations. Few of the simulations result in less than 2 °C of warming—near the low end of estimates by the Intergovernmental Panel on Climate Change (IPCC). Some simulations result in significantly more than the 4 °C, which is at the high end of the IPCC estimates. This pattern (statisticians call it a "right-skewed distribution") suggests that if carbon dioxide concentrations double, the probability of very large increases in temperature is greater than the probability of very small increases.

Climate sensitivity is the equilibrium temperature change in response to changes of the radiative forcing. Therefore, climate sensitivity depends on the initial climate state, but potentially can be accurately inferred from precise palaeoclimate data. Slow climate feedbacks, especially changes of ice sheet size and atmospheric CO_2, amplify the total Earth system sensitivity by an amount that depends on the time scale considered.

Although climate sensitivity is usually used in the context of radiative forcing by carbon dioxide (CO_2), it is thought of as a general property of the climate system: the change in surface air temperature (ΔT_s) following a unit change in radiative forcing (RF), and thus is expressed in units of °C/(W/m²). For this to be useful, the measure must be independent of the nature of the forcing (e.g. from greenhouse gases or solar variation); to first order this is indeed found to be so.

The climate sensitivity specifically due to CO_2 is often expressed as the temperature change in °C associated with a doubling of the concentration of carbon dioxide in Earth's atmosphere.

For coupled atmosphere-ocean global climate models (e.g. CMIP5) the climate sensitivity is an emergent property: it is not a model parameter, but rather a result of a combination of model physics and parameters. By contrast, simpler energy-balance models may have climate sensitivity as an explicit parameter.

$$\Delta T_s = \lambda \cdot RF$$

The terms represented in the equation relate radiative forcing (RF) to linear changes in global surface temperature change (ΔT_s) via the climate sensitivity λ.

It is also possible to estimate climate sensitivity from observations; however, this is difficult due to uncertainties in the forcing and temperature histories.

Equilibrium and Transient Climate Sensitivity

The equilibrium climate sensitivity (ECS) refers to the equilibrium change in global mean near-surface air temperature that would result from a sustained doubling of the atmospheric (equivalent) carbon dioxide concentration (ΔT_{x2}). As estimated by the IPCC Fifth Assessment Report (*AR5*) "there is *high confidence* that ECS is *extremely unlikely* less than 1°C and *medium confidence* that the ECS is *likely* between 1.5°C and 4.5°C and *very unlikely* greater than 6°C." This is a change from the IPCC Fourth Assessment Report (*AR4*), which said it was *likely to be in the range 2 to 4.5 °C with a best estimate of about 3 °C, and is very unlikely to be less than 1.5 °C. Values substantially higher than 4.5 °C cannot be excluded, but agreement of models with observations is not as good for those values.* The IPCC Third Assessment Report (*TAR*) said it was "likely to be in the range of 1.5 to 4.5 °C". Other estimates of climate sensitivity are discussed later on.

A model estimate of equilibrium sensitivity thus requires a very long model integration; fully equilibrating ocean temperatures requires integrations of thousands of model years. A measure requiring shorter integrations is the transient climate response (TCR) which is defined as the average temperature response over a twenty-year period centered at CO_2 doubling in a transient simulation with CO_2 increasing at 1% per year. The transient response is lower than the equilibrium sensitivity, due to the "inertia" of ocean heat uptake.

Over the 50–100 year timescale, the climate response to forcing is likely to follow the TCR; for considerations of climate stabilization, the ECS is more useful.

An estimate of the equilibrium climate sensitivity may be made from combining the transient climate sensitivity with the known properties of the ocean reservoirs and the surface heat fluxes; this is the effective climate sensitivity. This "may vary with forcing history and climate state".

A less commonly used concept, the Earth system sensitivity (ESS), can be defined which includes the effects of slower feedbacks, such as the albedo change from melting the large ice sheets that covered much of the northern hemisphere during the last glacial maximum. These extra feedbacks make the ESS larger than the ECS — possibly twice as large — but also mean that it may well not apply to current conditions.

Sensitivity to Carbon Dioxide Forcing

Climate sensitivity is often evaluated in terms of the change in equilibrium temperature due to radiative forcing due to the greenhouse effect. According to the Arrhenius relation, the radiative forcing (and hence the change in temperature) is proportional to the logarithm of the concentration of infrared-absorbing gasses in the atmosphere. Thus, the sensitivity of temperature to gasses in the atmosphere (most notably carbon dioxide) is often expressed in terms of the change in temperature per doubling of the concentration of the gas.

Radiative Forcing Due to Doubled CO_2

CO_2 climate sensitivity has a component directly due to radiative forcing by CO_2, and a further contribution arising from climate feedbacks, both positive and negative. "Without any feedbacks, a doubling of CO_2 (which amounts to a forcing of 3.7 W/m²) would result in 1 °C global warming, which is easy to calculate and is undisputed. The remaining uncertainty is due entirely to feedbacks in the system, namely, the water vapor feedback, the ice-albedo feedback, the cloud feedback, and the lapse rate feedback"; addition of these feedbacks leads to a value of the sensitivity to CO_2 doubling of approximately 3 °C ± 1.5 °C, which corresponds to a value of λ of 0.8 K/(W/m²).

In the earlier 1979 NAS report (p. 7), the radiative forcing due to doubled CO_2 is estimated to be 4 W/m², as calculated (for example) in Ramanathan et al. (1979). In 2001 the IPCC adopted the revised value of 3.7 W/m², the difference attributed to a "stratospheric temperature adjustment". More recently an intercomparison of radiative transfer codes (Collins et al., 2006) showed discrepancies among climate models and between climate models and more exact radiation codes in the forcing attributed to doubled CO_2 even in cloud-free sky; presumably the differences would be even greater if forcing were evaluated in the presence of clouds because of differences in the treatment of clouds in different models. Undoubtedly the difference in forcing attributed to doubled CO_2 in different climate models contributes to differences in apparent sensitivities of the models, although this effect is thought to be small relative to the intrinsic differences in sensitivities of the models themselves.

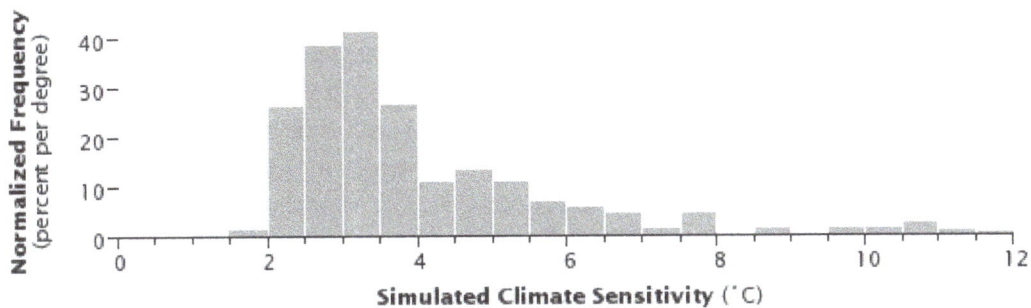

Frequency distribution of climate sensitivity, based on model simulations. Few of the simulations result in less than 2 °C of warming—near the low end of estimates by the Intergovernmental Panel on Climate Change (IPCC). Some simulations result in significantly more than the 4 °C, which is at the high end of the IPCC estimates. This pattern (statisticians call it a "right-skewed distribution") suggests that if carbon dioxide concentrations double, the probability of very large increases in temperature is greater than the probability of very small increases.

Consensus Estimates

A committee on anthropogenic global warming convened in 1979 by the National Academy of Sciences and chaired by Jule Charney estimated climate sensitivity to be 3 °C, plus or minus 1.5 °C. Only two sets of models were available; one, due to Syukuro Manabe, exhibited a climate sensitivity of 2 °C, the other, due to James E. Hansen, exhibited a climate sensitivity of 4 °C. "According to Manabe, Charney chose 0.5 °C as a not-unreasonable margin of error, subtracted it from Manabe's number, and added it to Hansen's. Thus was born the 1.5 °C-to-4.5 °C range of likely climate sensitivity that has appeared in every greenhouse assessment since…"

Chapter 4 of the "Charney report" compares the predictions of the models: "We conclude that the predictions … are basically consistent and mutually supporting. The differences in model results are relatively small and may be accounted for by differences in model characteristics and simplifying assumptions."

In 2008 climatologist Stefan Rahmstorf wrote, regarding the Charney report's original range of uncertainty: "At that time, this range was on very shaky ground. Since then, many vastly improved models have been developed by a number of climate research centers around the world. Current state-of-the-art climate models span a range of 2.6–4.1 °C, most clustering around 3 °C."

Intergovernmental Panel on Climate Change

The 1990 IPCC First Assessment Report estimated that equilibrium climate sensitivity to CO_2 doubling lay between 1.5 and 4.5 °C, with a "best guess in the light of current knowledge" of 2.5 °C. This used models with strongly simplified representations of the ocean dynamics. The IPCC supplementary report, 1992 which used full ocean GCMs nonetheless saw "no compelling reason to warrant changing" from this estimate and the IPCC Second Assessment Report found that "No strong reasons have emerged to change" these estimates, with much of the uncertainty attributed to cloud processes. As noted above, the IPCC TAR retained the likely range 1.5 to 4.5 °C.

Authors of the IPCC Fourth Assessment Report (Meehl *et al.*, 2007) stated that confidence in estimates of equilibrium climate sensitivity had increased substantially since the TAR. AR4's assessment was based on a combination of several independent lines of evidence, including observed climate change and the strength of known "feedbacks" simulated in general circulation models. IPCC authors concluded that the global mean equilibrium warming for doubling CO_2 (to a concentration of 560 ppmv), or equilibrium climate sensitivity, very likely is greater than 1.5 °C (2.7 °F) and likely to lie in the range 2 to 4.5 °C (4 to 8.1 °F), with a most likely value of about 3 °C (5 °F). For fundamental physical reasons, as well as data limitations, the IPCC states a climate sensitivity higher than 4.5 °C (8.1 °F) cannot be ruled out, but that agreement for these values with observations and "proxy" climate data is generally worse compared to values in the 2 to 4.5 °C (4 to 8.1 °F) range.

The TAR uses the word "likely" in a qualitative sense to describe the likelihood of the 1.5 to 4.5 °C range being correct. AR4, however, quantifies the probable range of climate sensitivity estimates:

- 2-4.5 °C is "likely", = greater than 66% chance of being correct

- less than 1.5 °C is "very unlikely" = less than 10%

The IPCC Fifth Assessment Report stated: Equilibrium climate sensitivity is likely in the range 1.5 °C to 4.5 °C (high confidence), extremely unlikely less than 1 °C (high confidence), and very unlikely greater than 6 °C (medium confidence).

These are Bayesian probabilities, which are based on an expert assessment of the available evidence.

Calculations of CO_2 Sensitivity from Observational Data

Sample Calculation using Industrial-age Data

Rahmstorf (2008) provides an informal example of how climate sensitivity might be estimated empirically, from which the following is modified. Denote the sensitivity, i.e. the equilibrium increase in global mean temperature including the effects of feedbacks due to a sustained forcing by doubled CO_2 (taken as 3.7 W/m²), as x °C. If Earth were to experience an equilibrium temperature change of ΔT (°C) due to a sustained forcing of ΔF (W/m²), then one might say that $x/(\Delta T)$ = (3.7 W/m²)/(ΔF), i.e. that $x = \Delta T * (3.7$ W/m²)/ΔF. The global temperature increase since the beginning of the industrial period (taken as 1750) is about 0.8 °C, and the radiative forcing due to CO_2 and other long-lived greenhouse gases (mainly methane, nitrous oxide, and chlorofluorocarbons) emitted since that time is about 2.6 W/m². Neglecting other forcings and considering the temperature increase to be an equilibrium increase would lead to a sensitivity of about 1.1 °C. However, ΔF also contains contributions due to solar activity (+0.3 W/m²), aerosols (-1 W/m²), ozone (0.3 W/m²) and other lesser influences, bringing the total forcing over the industrial period to 1.6 W/m² according to best estimate of the IPCC AR4, albeit with substantial uncertainty. Additionally the fact that the climate system is not at equilibrium must be accounted for; this is done by subtracting the planetary heat uptake rate H from the forcing; i.e., $x = \Delta T * (3.7$ W/m²)/$(\Delta F-H)$. Taking planetary heat uptake rate as the rate of ocean heat uptake, estimated by the IPCC AR4 as 0.2 W/m², yields a value for x of 2.1 °C. (All numbers are approximate and quite uncertain.)

Sample Calculation Using Ice-age Data

In 2008, Farley wrote: "… examine the change in temperature and solar forcing between glaciation (ice age) and interglacial (no ice age) periods. The change in temperature, revealed in ice core samples, is 5 °C, while the change in solar forcing is 7.1 W/m². The computed climate sensitivity is therefore 5/7.1 = 0.7 K(W/m²)⁻¹. We can use this empirically derived climate sensitivity to predict the temperature rise from a forcing of 4 W/m², arising from a doubling of the atmospheric CO_2 from pre-industrial levels. The result is a predicted temperature increase of 3 °C."

Based on analysis of uncertainties in total forcing, in Antarctic cooling, and in the ratio of global to Antarctic cooling of the last glacial maximum relative to the present, Ganopolski and Schneider von Deimling (2008) infer a range of 1.3 to 6.8 °C for climate sensitivity determined by this approach.

A lower figure was calculated in a 2011 *Science* paper by Schmittner *et al.*, who combined temperature reconstructions of the Last Glacial Maximum with climate model simulations to suggest a rate of global warming from doubling of atmospheric carbon dioxide of a median of 2.3 °C and uncertainty 1.7–2.6 °C (66% probability range), less than the earlier estimates of 2 to 4.5 °C as the 66% probability range. Schmittner et al. said their "results imply less probability of extreme climatic change than previously thought." Their work suggests that climate sensitivities >6 °C "cannot be reconciled with paleoclimatic and geologic evidence, and hence should be assigned near-zero probability."

Other Experimental Estimates

Idso (1998) calculated based on eight natural experiments a λ of 0.1 °C/(Wm⁻²) resulting in a climate sensitivity of only 0.4 °C for a doubling of the concentration of CO_2 in the atmosphere.

Andronova and Schlesinger (2001) found that the climate sensitivity could lie between 1 and 10 °C, with a 54 percent likelihood that it lies outside the IPCC range. The exact range depends on which factors are most important during the instrumental period: "At present, the most likely scenario is one that includes anthropogenic sulfate aerosol forcing but not solar variation. Although the value of the climate sensitivity in that case is most uncertain, there is a 70 percent chance that it exceeds the maximum IPCC value. This is not good news," said Schlesinger.

Forest, *et al.* (2002) using patterns of change and the MIT EMIC estimated a 95% confidence interval of 1.4–7.7 °C for the climate sensitivity, and a 30% probability that sensitivity was outside the 1.5 to 4.5 °C range.

Gregory, *et al.* (2002) estimated a lower bound of 1.6 °C by estimating the change in Earth's radiation budget and comparing it to the global warming observed over the 20th century.

Shaviv (2005) carried out a similar analysis for 6 different time scales, ranging from the 11-yr solar cycle to the climate variations over geological time scales. He found a typical sensitivity of 0.54 ± 0.12 K/(W m^{-2}) or 2.1 °C (ranging between 1.6 °C and 2.5 °C at 99% confidence) if there is no cosmic-ray climate connection, or a typical sensitivity of 0.35 ± 0.09 K/(W m^{-2}) or 1.3 °C (between 1.0 °C and 1.7 °C at 99% confidence), if the cosmic-ray climate link is real. (Note Shaviv quotes a radiative forcing equivalent of 3.8 Wm^{-2}. [ΔT_{x2}=3.8 Wm^{-2} λ].)

Frame, *et al.* (2005) noted that the range of the confidence limits is dependent on the nature of the prior assumptions made.

Annan and Hargreaves (2006) presented an estimate that resulted from combining prior estimates based on analyses of paleoclimate, responses to volcanic eruptions, and the temperature change in response to forcings over the twentieth century. They also introduced a triad notation (L, C, H) to convey the probability distribution function (pdf) of the sensitivity, where the central value C indicates the maximum likelihood estimate in degrees Celsius and the outer values L and H represent the limits of the 95% confidence interval for a pdf, or 95% of the area under the curve for a likelihood function. In this notation their estimate of sensitivity was (1.7, 2.9, 4.9) °C.

Forster and Gregory (2006) presented a new independent estimate based on the slope of a plot of calculated greenhouse gas forcing minus top-of-atmosphere energy imbalance, as measured by satellite borne radiometers, versus global mean surface temperature. In the triad notation of Annan and Hargreaves their estimate of sensitivity was (1.0, 1.6, 4.1) °C.

Royer, *et al.* (2007) determined climate sensitivity within a major part of the Phanerozoic. The range of values—1.5 °C minimum, 2.8 °C best estimate, and 6.2 °C maximum—is, given various uncertainties, consistent with sensitivities of current climate models and with other determinations.

Lindzen and Choi (2011) find the equilibrium climate sensitivity to be 0.7 C, implying a negative feedback of clouds.

Skeie et al. (2013) use the Bayesian analysis of the OHC data and conclude that the equilibrium climate sensitivity is 1.8 C, far lower than previous best estimate relied upon by the IPCC.

Aldrin et al. (2012) use a simple deterministic climate model, modelling yearly hemispheric surface temperature and global ocean heat content as a function of historical radiative forcing and combine it with an empirical, stochastic model. By using a Bayesian framework they estimate the equilibrium climate sensitivity to be 1.98 C.

Lewis (2013) estimates by using the Bayesian framework that the equilibrium climate sensitivity is 1.6 K, with the likely range (90% confidence level) 1.2-2.2 K.

ScienceDaily reported on a study by Fasullo and Trenberth (2012), who tested model estimates of climate sensitivity based on their ability to reproduce observed relative humidity in the tropics and subtropics. The best performing models tended to project relatively high climate sensitivities, of around 4 °C.

Previdi et al. 2013 reviewed the 2×CO_2 Earth system sensitivity, and concluded it is higher if the ice sheet and the vegetation albedo feedback is included in addition to the fast feedbacks, being ~4–6 °C, and higher still if climate–GHG feedbacks are also included.

Lewis and Curry (2014) estimated that equilibrium climate sensitivity was 1.64 °C, based on the 1750-2011 time series and "the uncertainty ranges for forcing components" in the IPCC's Fifth Assessment Report.

Literature Reviews

A literature review by Knutti and Hegerl (2008) concluded that "various observations favour a climate sensitivity value of about 3 °C, with a likely range of about 2-4.5 °C. However, the physics of the response and uncertainties in forcing lead to difficulties in ruling out higher values."

Radiative Forcing Functions

A number of different inputs can give rise to radiative forcing. In addition to the downwelling radiation due to the greenhouse effect, the IPCC First Scientific Assessment Report listed solar radiation variability due to orbital changes, variability due to changes in solar irradiance, direct aerosol effects (*e.g.*, changes in albedo due to cloud cover), indirect aerosol effects, and surface characteristics.

Sensitivity to Solar Forcing

Solar irradiance is about 0.9 W/m² brighter during solar maximum than during solar minimum. Analysis by Camp and Tung shows that this correlates with a variation of ±0.1 °C in measured average global temperature between the peak and minimum of the 11-year solar cycle. From this data (incorporating the Earth's albedo and the fact that the solar absorption cross-section is 1/4 of the surface area of the Earth), Tung, Zhou and Camp (2008) derive a transient sensitivity value of 0.69 to 0.97 °C/(W/m²). This would correspond to a transient climate sensitivity to carbon dioxide doubling of 2.5 to 3.6 K, similar to the range of the current scientific consensus. However, they note that this is the transient response to a forcing with an 11-year cycle; due to lag effects, they estimate the equilibrium response to forcing would be about 1.5 times as high.

Greenhouse Effect

The greenhouse effect is the process by which radiation from a planet's atmosphere warms the planet's surface to a temperature above what it would be without its atmosphere.

If a planet's atmosphere contains radiatively active gases (i.e., greenhouse gases) the atmosphere will radiate energy in all directions. Part of this radiation is directed towards the surface, warming it. The downward component of this radiation – that is, the strength of the greenhouse effect – will depend on the atmosphere's temperature and on the amount of greenhouse gases that the atmosphere contains.

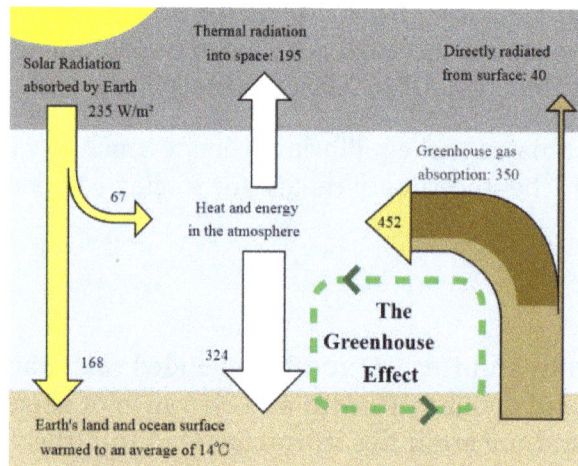

A representation of the exchanges of energy between the source (the Sun), Earth's surface, the Earth's atmosphere, and the ultimate sink outer space. The ability of the atmosphere to capture and recycle energy emitted by Earth's surface is the defining characteristic of the greenhouse effect.

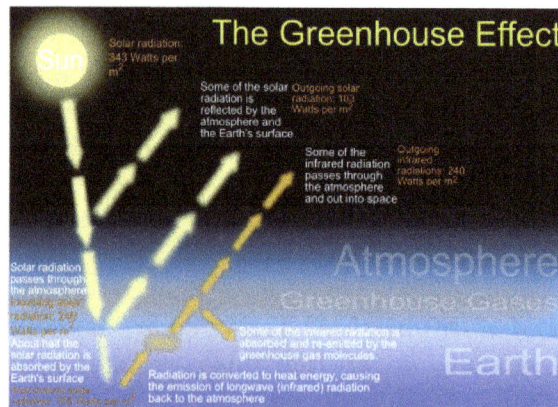

Another diagram of the greenhouse effect

On Earth, the atmosphere is warmed by absorption of infrared thermal radiation from the underlying surface, absorption of shorter wavelength radiant energy from the sun, and convective heat fluxes from the surface. Greenhouse gases in the atmosphere radiate energy, some of which is directed to the surface and lower atmosphere. The mechanism that produces this difference between the actual surface temperature and the effective temperature is due to the atmosphere and is known as the greenhouse effect.

Earth's natural greenhouse effect is critical to supporting life. Human activities, primarily the burning of fossil fuels and clearing of forests, have intensified the natural greenhouse effect, causing global warming.

The mechanism is named after a faulty analogy with the effect of solar radiation passing through glass and warming a greenhouse. The way a greenhouse retains heat is fundamentally different, as a greenhouse works by reducing airflow and retaining warm air inside the structure.

History

The existence of the greenhouse effect was argued for by Joseph Fourier in 1824. The argument and the evidence was further strengthened by Claude Pouillet in 1827 and 1838, and reasoned from experimental observations by John Tyndall in 1859. The effect was more fully quantified by Svante Arrhenius in 1896. However, the term "greenhouse" was not used to refer to this effect by any of these scientists; the term was first used in this way by Nils Gustaf Ekholm in 1901.

In 1917 Alexander Graham Bell wrote "[The unchecked burning of fossil fuels] would have a sort of greenhouse effect", and "The net result is the greenhouse becomes a sort of hot-house." Bell went on to also advocate the use of alternate energy sources, such as solar energy.

Mechanism

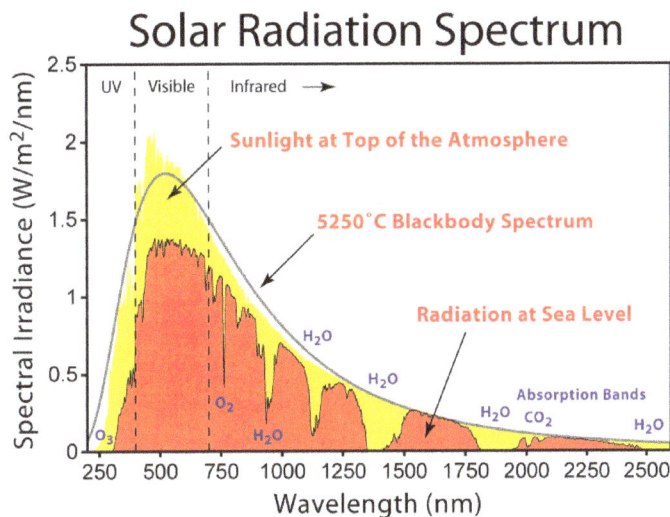

The solar radiation spectrum for direct light at both the top of Earth's atmosphere and at sea level

Earth receives energy from the Sun in the form of ultraviolet, visible, and near-infrared radiation. Of the total amount of solar energy available at the top of the atmosphere, about 26% is reflected to space by the atmosphere and clouds and 19% is absorbed by the atmosphere and clouds. Most of the remaining energy is absorbed at the surface of Earth. Because the Earth's surface is colder than the photosphere of the Sun, it radiates at wavelengths that are much longer than the wavelengths that were absorbed. Most of this thermal radiation is absorbed by the atmosphere, thereby warming it. In addition to the absorption of solar and thermal radiation, the atmosphere further gains heat by sensible and latent heat fluxes from the surface. The atmosphere radiates energy both upwards and downwards; the part radiated downwards is absorbed by the surface of Earth. This leads to a higher equilibrium temperature than if the atmosphere were absent.

An ideal thermally conductive blackbody at the same distance from the Sun as Earth would have a temperature of about 5.3 °C. However, because Earth reflects about 30% of the incoming sunlight, this idealized planet's effective temperature (the temperature of a blackbody that would emit the same amount of radiation) would be about −18 °C. The surface temperature of this hypothetical planet is 33 °C below Earth's actual surface temperature of approximately 14 °C.

The basic mechanism can be qualified in a number of ways, none of which affect the fundamental process. The atmosphere near the surface is largely opaque to thermal radiation (with important exceptions for "window" bands), and most heat loss from the surface is by sensible heat and latent heat transport. Radiative energy losses become increasingly important higher in the atmosphere, largely because of the decreasing concentration of water vapor, an important greenhouse gas. It is more realistic to think of the greenhouse effect as applying to a "surface" in the mid-troposphere, which is effectively coupled to the surface by a lapse rate. The simple picture also assumes a steady state, but in the real world there are variations due to the diurnal cycle as well as the seasonal cycle and weather disturbances. Solar heating only applies during daytime. During the night, the atmosphere cools somewhat, but not greatly, because its emissivity is low. Diurnal temperature changes decrease with height in the atmosphere.

Within the region where radiative effects are important, the description given by the idealized greenhouse model becomes realistic. Earth's surface, warmed to a temperature around 255 K, radiates long-wavelength, infrared heat in the range of 4–100 μm. At these wavelengths, greenhouse gases that were largely transparent to incoming solar radiation are more absorbent. Each layer of atmosphere with greenhouses gases absorbs some of the heat being radiated upwards from lower layers. It reradiates in all directions, both upwards and downwards; in equilibrium (by definition) the same amount as it has absorbed. This results in more warmth below. Increasing the concentration of the gases increases the amount of absorption and reradiation, and thereby further warms the layers and ultimately the surface below.

Greenhouse gases—including most diatomic gases with two different atoms (such as carbon monoxide, CO) and all gases with three or more atoms—are able to absorb and emit infrared radiation. Though more than 99% of the dry atmosphere is IR transparent (because the main constituents—N_2, O_2, and Ar—are not able to directly absorb or emit infrared radiation), intermolecular collisions cause the energy absorbed and emitted by the greenhouse gases to be shared with the other, non-IR-active, gases.

Greenhouse Gases

Atmospheric gases only absorb some wavelengths of energy but are transparent to others. The absorption patterns of water vapor (blue peaks) and carbon dioxide (pink peaks) overlap in some wavelengths. Carbon dioxide is not as strong a greenhouse gas as water vapor, but it absorbs energy in wavelengths (12-15 micrometers) that water vapor does not, partially closing the "window" through which heat radiated by the surface would normally escape to space. (Illustration NASA, Robert Rohde)

By their percentage contribution to the greenhouse effect on Earth the four major gases are:

- water vapor, 36–70%

- carbon dioxide, 9–26%

- methane, 4–9%

- ozone, 3–7%

It is not physically realistic to assign a specific percentage to each gas because the absorption and emission bands of the gases overlap (hence the ranges given above). The major non-gas contributor to Earth's greenhouse effect, clouds, also absorb and emit infrared radiation and thus have an effect on the radiative properties of the atmosphere.

Role in Climate Change

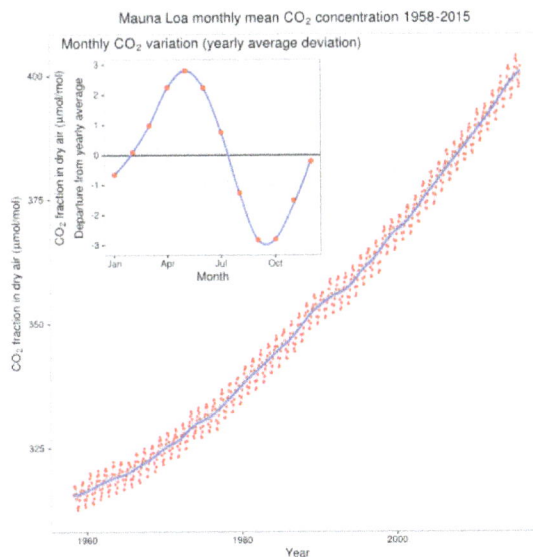

The Keeling Curve of atmospheric CO_2 concentrations measured at Mauna Loa Observatory.

Strengthening of the greenhouse effect through human activities is known as the enhanced (or anthropogenic) greenhouse effect. This increase in radiative forcing from human activity is attributable mainly to increased atmospheric carbon dioxide levels. According to the latest Assessment Report from the Intergovernmental Panel on Climate Change, "*atmospheric concentrations of carbon dioxide, methane and nitrous oxide are unprecedented in at least the last 800,000 years. Their effects, together with those of other anthropogenic drivers, have been detected throughout the climate system and are extremely likely to have been the dominant cause of the observed warming since the mid-20th century*".

CO_2 is produced by fossil fuel burning and other activities such as cement production and tropical deforestation. Measurements of CO_2 from the Mauna Loa observatory show that concentrations have increased from about 313 parts per million (ppm) in 1960 to about 389 ppm in 2010. It reached the 400 ppm milestone on May 9, 2013. The current observed amount of CO_2 exceeds the geological record maxima (~300 ppm) from ice core data. The effect of combustion-produced carbon dioxide on the global climate, a special case of the greenhouse effect first described in 1896 by Svante Arrhenius, has also been called the Callendar effect.

Over the past 800,000 years, ice core data shows that carbon dioxide has varied from values as low as 180 ppm to the pre-industrial level of 270 ppm. Paleoclimatologists consider variations in carbon dioxide concentration to be a fundamental factor influencing climate variations over this time scale.

Real Greenhouses

A modern Greenhouse in RHS Wisley

The "greenhouse effect" of the atmosphere is named by analogy to greenhouses which become warmer in sunlight. The explanation given in most sources for the warmer temperature in an actual greenhouse is that incident solar radiation in the visible, long-wavelength ultraviolet, and short-wavelength infrared range of the spectrum passes through the glass roof and walls and is absorbed by the floor, earth, and contents, which become warmer and re-emit the energy as longer-wavelength infrared radiation. Glass and other materials used for greenhouse walls do not transmit infrared radiation, so the infrared cannot escape via radiative transfer. As the structure is not open to the atmosphere, heat also cannot escape via convection, so the temperature inside the greenhouse rises. The greenhouse effect, due to infrared-opaque "greenhouse gases" including carbon dioxide and methane instead of glass, also affects Earth as a whole; there is no convective cooling because no significant amount of air escapes from Earth.

However the mechanism by which the atmosphere retains heat—the "greenhouse effect"—is different; a greenhouse is not primarily warmed by the "greenhouse effect". A greenhouse works primarily by allowing sunlight to warm surfaces inside the structure, but then preventing absorbed heat from leaving the structure through convection. The "greenhouse effect" heats Earth because greenhouse gases absorb outgoing radiative energy, heating the atmosphere which then emits radiative energy with some of it going back towards Earth.

A greenhouse is built of any material that passes sunlight, usually glass, or plastic. It mainly warms up because the sun warms the ground and contents inside, which then warms the air in the greenhouse. The air continues to heat up because it is confined within the greenhouse, unlike the environment outside the greenhouse where warm air near the surface rises and mixes with cooler air aloft. This can be demonstrated by opening a small window near the roof of a greenhouse: the temperature will drop considerably. It was demonstrated experimentally (R. W. Wood, 1909) that a "greenhouse" with a cover of rock salt (which is transparent to infrared) heats up an enclosure similarly to one with a glass cover. Thus greenhouses work primarily by preventing convective cooling.

More recent quantitative studies suggest that the effect of infrared radiative cooling is not negligibly small, and may have economic implications in a heated greenhouse. Analysis of issues of near-infrared radiation in a greenhouse with screens of a high coefficient of reflection concluded that installation of such screens reduced heat demand by about 8%, and application of dyes to transparent surfaces was suggested. Composite less-reflective glass, or less effective but cheaper anti-reflective coated simple glass, also produced savings.

Bodies other than Earth

In the Solar System, there also greenhouse effects on Mars, Venus, and Titan. The greenhouse effect on Venus is particularly large because its dense atmosphere consisting mainly of carbon dioxide. Titan has an anti-greenhouse effect, in that its atmosphere absorbs solar radiation but is relatively transparent to infrared radiation. Pluto is also colder than would be expected, because evaporation of nitrogen cools it.

A runaway greenhouse effect occurs if positive feedbacks lead to the evaporation of all greenhouse gases into the atmosphere. A runaway greenhouse effect involving carbon dioxide and water vapor is thought to have occurred on Venus.

Tectonic–climatic Interaction

Tectonic–climatic interaction is the interrelationship between tectonic processes and the climate system. The tectonic processes in question include orogenesis, volcanism, and erosion, while relevant climatic processes include atmospheric circulation, orographic lift, monsoon circulation and the rain shadow effect. As the geological record of past climate changes over millions of years is sparse and poorly resolved, many questions remain unresolved regarding the nature of tectonic-climate interaction, although it is an area of active research by geologists and palaeoclimatologists.

Orographic Controls on Climate

Depending on the vertical and horizontal magnitude of a mountain range, it has the potential to have strong effects on global and regional climate patterns and processes including: deflection of atmospheric circulation, creation of orographic lift, altering monsoon circulation, and causing the rain shadow effect.

One example of an elevated terrain and its effect on climate occurs in the Southeast Asian Himalayas, the world's highest mountain system. A range of this size has the ability to influence geographic temperature, precipitation, and wind. Theories suggest that the uplift of the Tibetan Plateau has resulted in stronger deflections of the atmospheric jet stream, a heavier monsoonal circulation, increased rainfall on the front slopes, greater rates of chemical weathering, and thus lower atmospheric CO_2 concentrations. It is possible that the spatial magnitude of this range is so great that it creates a regional monsoon circulation in addition to disrupting hemispheric-scale atmospheric circulation.

Simple illustration of the *rain shadow effect*

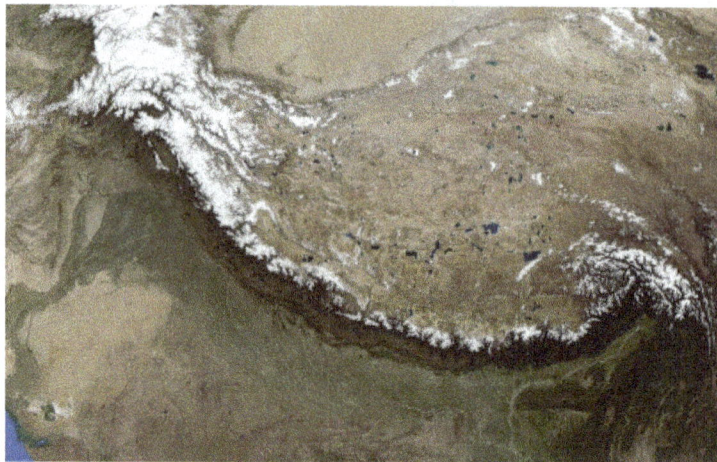

Example of the *rain shadow effect* in the *Himalayas*

The monsoon season in Southeast Asia occurs due to the Asian continent becoming warmer than the surrounding oceans during the summer; as a low-pressure cell is created above the continents, a high-pressure cell forms over the cooler ocean, causing advection of moist air, creating heavy precipitation from Africa to Southeast Asia. However, the intensity of the rainfall over Southeast Asia is greater than the African monsoon, which can be attributed to the awesome size of the Asian continent compared to the African continent and the presence of a vast mountain system. This not only affects the climate of Southeast Asia, but modifies the climate in neighboring areas such as Siberia, central Asia, the Middle East, and the Mediterranean basin as well. To test this a model was created that changed only the topography of current landmasses, which resulted in correlations between the model and global fluctuations in precipitation and temperature over the past 40 Myr. interpreted by scientists.

It is commonly agreed upon that global climate fluctuations are strongly dictated by the presence or absence of greenhouse gases in the atmosphere and carbon dioxide (CO_2) is typically considered the most significant greenhouse gas. Observations infer that large uplifts of mountain ranges globally result in higher chemical erosion rates, thus lowering the volume of CO_2 in the atmosphere as well as causing global cooling. This occurs because in regions of higher elevation there are higher rates of mechanical erosion (i.e. gravity, fluvial processes) and there is constant exposure and availability of materials available for chemical weathering. The following is a simplified equation describing the consumption of CO_2 during chemical weathering of silicates:

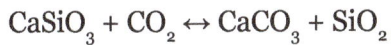

$$CaSiO_3 + CO_2 \leftrightarrow CaCO_3 + SiO_2$$

From this equation, it is inferred that carbon dioxide is consumed during chemical weathering and thus lower concentrations of the gas will be present in the atmosphere as long as chemical weathering rates are high enough.

Climate-driven Tectonism

There are scientists who reject that uplift is the sole cause of climate change and are in favor of uplift as a result of climate change. Some geologists theorize that a cooler and stormier climate (such as glaciations and increased precipitation) can give a landscape a younger appearance such as incision of high terrains and increased erosion rates. Glaciers are a powerful eroding agent with the ability to incise and carve deep valleys and when rapid erosion of the earth's surface occurs, especially in an area of limited relief, it is possible for isostatic rebound to occur, creating high peaks and deep valleys. A lack of glaciation or precipitation can cause an increase in erosion, but can vary between localities. It is possible to create erosion in the absence of precipitation because there would be a decrease in vegetation, which typically acts as a protective cover for the bedrock.

Peaks and valleys of the *Torres del Paine* range of the *Andes* in *Chile*

Models also suggest that certain topographic features of the Himalayan and Andes region are determined by an erosional/climatic interaction as opposed to tectonism. These models reveal a correlation between regional precipitation and a maximum topographic limit at the plateau margin. In the southern Andes where there is relatively low precipitation and denudation rates, there is no real extreme topography present at the plateau margin while in the north there are higher rates of precipitation and the presence of extreme topography.

Another interesting theory comes from an investigation of the uplift of the Andes during the Cenozoic. Some scientists hypothesize that the tectonic processes of plate subduction and mountain building are products of erosion and sedimentation. When there is an arid climate influenced by the rain shadow effect in a mountainous region, sediment supply to the trench can be reduced or even cut off. These sediments are thought to act as lubricants at the plate interface and this reduction increases the shear stress present at the interface that is large enough to support the high Andes.

Volcanism

Introduction

Around the world, dotting the map are volcanoes of all shapes and sizes. Lining the landmass around the Pacific Ocean are the well-known volcanoes of the Pacific Ring of Fire. From the Aleutian Islands to the Andes Mountains in Chile, these volcanoes have sculpted their local and regional environments. Aside from admiring their majestic beauty, one might wonder how these geologic wonders work and what role they play in changing the landscape and atmosphere. Principally, volcanoes are geologic features that exude magmatic material from below Earth's surface onto the surface. Upon reaching the surface, the term "magma" disappears and "lava" becomes the common nomenclature. This lava cools and forms igneous rock. By examining igneous rocks, it is possible to derive a chain of events that led from the original melt of the magma to the crystallization of the lava at Earth's surface. By examining igneous rocks, it is possible to postulate evidence for volcanic outgassing, which is known to alter atmospheric chemistry. This alteration of atmospheric chemistry changes climate cycles both globally and locally.

Fundamentals of Igneous Rock and Magmatic Gas Formation

Magmas are the starting point for the creation of a volcano. In order to understand volcanism, it is critical to understand the processes that form volcanoes. Magmas are created by keeping temperature, pressure, and composition (known as P-T-X) in the realm of melt conditions. The pressure and temperature for melts are understood by knowing the chemistry of the melt. To keep magma in a melt condition, a change in one variable will result in the change of another variable in order to maintain equilibrium (i.e. Le Chatlier's Principle). The production of magma is accomplished in multiple ways: 1) subduction of oceanic crust, 2) creation of a hot spot from a mantle plume, and 3) divergence of oceanic or continental plates. The subduction of oceanic crust produces a magmatic melt usually at great depth. Yellowstone National Park is a hot spot located within the center of a continent. Divergence of continental plates (i.e. the Atlantic Mid-Ocean ridge complex) creates magmas very near the surface of the Earth. A plume of heat from the mantle will melt rocks, creating a hot spot, which can be located at any depth in the crust. Hot spots in oceanic crust develop different magmatic plumbing systems based on plate velocities. Hawaii and the Madeira Archipelago (off the West coast of Africa) are examples of volcanic complexes with two different plumbing systems. Because islands such as Hawaii move more quickly than Madeira, the layered rocks at Hawaii have a different chemistry than those at Madeira. The layers beneath Hawaii and Madeira are different because the magma produced underground at these locations rests for different amounts of time. The longer the amount of time magma will rest underground, the warmer the host rocks become. Fractionation of crystals from melt is partially driven by heat; therefore, the igneous rock produced will vary from an insulated host to a non-insulated host. Each of these avenues of magmatic creation develops different igneous rocks and, thus, various P-T-X histories. Definitions and other geologic explanations of igneous systems are explained in Loren A. Raymond's *Petrology* text.

In order to understand the creation of igneous rocks from a melt, it is fundamental to understand the concepts produced by Drs. Norman Bowen and Frank Tuttle from the $NaAlSiO_4$-$KAlSiO_4$-SiO_2-H_2O system. Tuttle and Bowen accomplished their work by using experimental petrologic laboratories that produce synthetic igneous materials from mixes of reagents. Observations from these experiments indicate that as a melt cools, it will produce derivative magmas and igneous rock. Following Bowen's research, the magma will crystallize a mafic igneous rock prior to a felsic igneous rock. As this crystallization process occurs in nature, pressure and temperature decrease, which changes the composition of the melt along various stages of the process. This constantly changing chemical environment alters the final composition that reaches the Earth's surface.

The evolution of magmatic gases depends on the P-T-X history of the magma. These factors include the composition of assimilated materials and composition of parent rock. Gases develop in magma through two different processes: first and second boiling. First boiling is defined as a decrease in confining pressure below the vapor pressure of the melt. Second boiling is defined as an increase in vapor pressure due to crystallization of the melt. In both cases, gas bubbles exsolve in the melt and aid the ascent of the magma towards the surface. As the magma ascends towards the surface, the temperature and confining pressure decrease. A decrease in temperature and confining pressure will allow an increase in crystallization and vapor pressure of the dissolved gas. Depending on the composition of the melt, this ascent can be either slow or fast. Felsic magmas are very viscous and travel to the surface of the Earth slower than mafic melts whose silica levels are lower. The amount of gas available to be exsolved and the concentrations of gases in the melt also control ascension of the magma. If the melt contains enough dissolved gas, the rate of exsolution will determine the magmas rate of ascension. Mafic melts contain low levels of dissolved gases whereas felsic melts contain high levels of dissolved gases. The rate of eruption for volcanoes of different compositions is not the controlling factor of gas emission into the atmosphere. The amount of gas delivered by an eruption is controlled by the origin of the magma, the crustal path the magma travels through, and several factors dealing with P-T-x at the Earth's surface. When felsic melts reach the surface of the Earth, they are generally very explosive (i.e. Mount St. Helens). Mafic melts generally flow over the surface of the Earth and form layers (i.e. Columbia River Basalt). Magma development under continental crust develops a different type of volcano than magmas that are generated under oceanic crust. Subduction zones produce volcanic island arcs (such as the Aleutian Islands, Alaska) and non-arc volcanism (such as Chile and California). Typically, arc volcanism is more explosive than non-arc volcanism due to the concentrations and amounts of gasses withheld in the magma underground.

Fluid inclusion analysis from fluids trapped in minerals can show a path of volatile evolution in volcanic rocks. Isotopic analyses and interpretation of degassing scenarios are required in order to derive the origin of magmatic volatiles. When gas bubbles accumulate in a melt that is crystallizing, they create a vesicular texture. Vesicules are created by super cooling a melt while gases are present. Because the rock crystallized very quickly while in the Earth's atmosphere, it is possible to examine some igneous rocks for fluids trapped in vesicles. By examining many different inclusions, it is possible to detect crustal assimilation and depressurization that account for volatile release.

Methods of Characterizing Igneous Rocks

The methods by which petrologists examine igneous rocks and synthetically produced materials are optical petrography, X-ray diffraction (XRD), electron probe microanalysis (EPMA), laser ablation inductively coupled mass spectrometry (LA-ICP-MS), and many others. Methods such as optical petrography aid the researcher in understanding various textures of igneous rocks and, also, the mineralogical composition of the rock. XRD methods define the mineralogical constituents of the rock being tested; therefore, composition is only known based on the mineralogical composition discovered using this method. EPMA reveals textural features of the rock on the micron level. It also reveals a composition of the rock based on elemental abundance. For information about fluids trapped in an igneous rock, LA-ICP-MS could be used. This is accomplished by finding rocks with small pockets of fluid or vapor, acquiring the fluid or vapor, and testing the fluid or vapor for various elements and isotopes.

Volcanic Emissions and Effects

While most volcanoes emit some mixture of the same few gasses, each volcano's emissions contain different ratios of those gasses. Water vapor (H_2O) is the predominant gas molecule produced, closely followed by carbon dioxide (CO_2) and sulfur dioxide (SO_2), all of which can function as greenhouse gasses. A few unique volcanoes release more unusual compounds. For example, mud volcanoes in Romania belch out much more methane gas than H_2O, CO_2, or SO_2 –95–98% methane (CH_4), 1.5–2.3% CO_2, and trace amounts of hydrogen and helium gas. To measure volcanic gases directly, scientists commonly use flasks and funnels to capture samples directly from volcanic vents or fumaroles. The advantage of direct measurement is the ability to evaluate trace levels in the gaseous composition. Volcanic gasses can be indirectly measured using Total Ozone Mapping Spectrometry (TOMS), a satellite-remote sensing tool which evaluates SO_2 clouds in the atmosphere. TOMS' disadvantage is that its high detection limit can only measure large amounts of exuded gases, such as those emitted by an eruption with an Volcanic Explosivity Index (VEI) of 3, on a logarithmic scale of 0 to 7.

Sulfur ejection from volcanoes has a tremendous impact environmental impact, and is important to consider when studying the large-scale effects of volcanism. Volcanoes are the primary source of the sulfur (in the form of SO_2) that ends up in the stratosphere, where it then reacts with OH radicals to form sulfuric acid (H_2SO_4). When the sulfuric acid molecules either spontaneously nucleate or condense on existing aerosols, they can grow large enough to form nuclei for raindrops and precipitate as acid rain. Rain containing elevated concentrations of SO_2 kills vegetation, which then reduces the ability of the area's biomass to absorb CO_2 from the air. It also creates a reducing environment in streams, lakes, and groundwater. Because of its high reactivity with other molecules, increased sulfur concentrations in the atmosphere can lead to ozone depletion and start a positive warming feedback.

Volcanoes with a felsic melt composition produce extremely explosive eruptions that can inject massive quantities of dust and aerosols high into the atmosphere. These particulate emissions are potent climate forcing agents, and can provoke a wide variety of responses including warming, cooling, and rainwater acidification. The climatic response depends on the altitude of the dust cloud as well as the size and composition of the dust. Some volcanic silicates cooled extremely quickly, created a glassy texture; their dark color and reflective nature absorb some radiation and reflect the rest. Such volcanic material injected into the stratosphere blocks solar radiation, heating that layer of the atmosphere and cooling the area beneath it. Wind patterns can distribute the dust over vast geographic regions; for example, the 1815 eruption of Tambora in Indonesia produced so much dust that a cooling of 1 degree Celsius was noted as far away as New England, and lasted for several months. Europeans and Americans called its effect "the year without a summer".

Volcanic emissions contain trace amounts of heavy metals, which can affect the hydrosphere when they are injected into the lower reaches of the atmosphere. When large quantities of these emissions are concentrated into a small area they can damage ecosystems, negatively affect agriculture, and pollute water sources.

Materials being emitted from volcanoes typically carry heavy metals in the trace level. When large quantities of these emissions are collected into a small area, the contamination effects become paramount.

The short-term (months-to-years) impacts of volcanism on the atmosphere, climate and environment are strongly controlled by location, timing, flux, magnitude and emission height of sulfur gases. Episodic explosive eruptions represent the principal perturbation to stratospheric aerosol (though the atmospheric effects of sulfur degassing associated with continental flood basalts might well be more profound). In the troposphere, the picture is less clear but a significant part of the global tropospheric sulfate burden may be volcanogenic. Sulfate aerosol influences the Earth's radiation budget by scattering and absorption of shortwave and long-wave radiation, and by acting as cloud condensation nuclei. When they are brought to the boundary layer and Earth's surface, clouds containing volcanic sulfur in both gaseous and aerosol phases can result in profound environmental and health impacts.

Examples of the environmental and health impacts are agricultural loss due to acid rain and particulate shading, damage to ecosystems, and pollution in the hydrosphere. Intensity of a volcanic eruption is a variable controlling the altitude and effect of ejected material. Though larger eruptions occur less often than smaller eruptions, larger eruptions still deliver more particulate matter into the atmosphere. This year round behavior of emitted material yields mild effects on the atmosphere in comparison to larger eruptions. Over time, changes in the composition of smaller scale eruptions yields changes to atmospheric cycles and the global climate. Larger scale eruptions cause changes to the atmosphere immediately, which in turn leads to climatic changes in the immediate vicinity. The larger the volcanic expulsion, the higher the altitude achieved by the ejected silicate materials. Higher altitude injections are caused by larger intensity eruptions. Larger eruptions do not emit as much, on average, as smaller eruptions. This is related to the return period of the eruptions and the amount of ejected material per eruption. "The injection height of sulfur into the atmosphere represents another important determinant of climate impact. More intense eruptions, i.e., those with higher magma discharge rates, are more likely to loft the reactive sulfur gases into the stratosphere where they can generate climatically effective aerosol."

Eruption intensity of a volcano is not the only factor controlling the altitude of particles. The climate surrounding the volcano constrains the impact of the eruption. Models of eruptions that treat climatic variables as controls and hold eruption intensity constant predict particulate emissions, such as volcanic ash and other pyroclastic debris ejected into the atmosphere, in the tropics to reach higher altitudes than eruptions in arid or polar areas. Some of these climatic variables include humidity, aridity, winds, and atmospheric stability. The observation made by the model matches what is seen in nature: volcanoes in tropical climates have greater eruption heights than those in the poles. If there were a widening of the tropics, the number of volcanoes able to produce higher altitude emissions into the atmosphere would increase. Effects on the climate from the increase in airborne silicate material would be substantial because the height of these tropical eruptions will become more prominent with a widening of the tropics leading to more risks such as cooling, pollution, and aircraft disturbances.

The location of a volcano strongly influences the geographic distribution of atmospheric heating and the development of planetary waves that affect air circulation (especially in the northern hemisphere). Another relevant factor is that the height of the tropopause varies with latitude—at the tropics it is around 16–17 km above sea level but descends to 10–11 km at high latitudes. In general terms, an explosive eruption requires a greater intensity (magma discharge rate) to cross the tropopause in the tropics than at mid to polar latitudes. However, there are two factors that limit this effect. The first is that a high-latitude eruption will have a more limited effect than a low-latitude one because further from the tropics there is less solar energy to intercept. Secondly, atmospheric circulation works in a way to limit the effects of high latitude eruptions. A tropical eruption that pumps aerosol into the stratosphere results in localized heating. This increases the temperature difference in the middle atmosphere between the equator and high latitudes, and thereby enhances meridional air flows that spread aerosol into both hemispheres, promoting climate forcing at a worldwide scale. In contrast, volcanic aerosol injected into the stratosphere from high latitude volcanoes will tend to have the opposite effect on the temperature gradient, acting to stagnate meridional air flow. Very little, if any, of the stratospheric aerosol formed as a result of eruption of a high latitude volcano will reach the opposing hemisphere.

Interaction between Glaciation and Volcanoes

Volcanoes do not only affect the climate, they are affected by the climate. During times of glaciation, volcanic processes slow down. Glacial growth is promoted when summer heat is weak and winter cold is enhanced and when glaciers grow larger, they get heavier. This excess weight causes a reverse effect on the magma chamber's ability to produce a volcano. Thermodynamically, magma will dissolve gases more readily when the confining pressure on the magma is greater than the vapor pressure of the dissolved components. Glacial buildup typically occurs at high elevations, which are also the home to most continental volcanoes. Buildup of ice can cause a magma chamber to fail and crystallize underground. The cause of magma chamber failure occurs when the pressure of ice pressing down on Earth is greater than the pressure being exerted on the magma chamber from heat convection in the mantle. Ice core data from glaciers provides insight into past climate. "Oxygen isotopes and the calcium ion record are essential indicators of climatic variability, while peaks in sulfate ions (SO_4) and in electrical conductivity of the ice indicate volcanic aerosol fallout." As seen in ice cores, volcanic eruptions in the tropics and southern hemisphere are not recorded in the Greenland Ice sheets. Fallout from tropical eruptions can be seen at both poles though this takes nearly two years and consists of only sulfuric precipitation. "One of the striking revelations of the ice core record is the evidence for numerous great eruptions, which have not otherwise been recognized in tephra records. One caveat to the approach is that although the dating of the ice core by counting of seasonal layers is fairly robust, it is not fail-safe. The greater the depth from which the core is retrieved, the more likely it is to have suffered deformation Prevailing winds and atmospheric chemistry play a large role in moving volcanic volatiles from their source to their final locations at the surface or in the atmosphere."

Cretaceous Climate

During the Cretaceous, Earth experienced an unusual warming trend. Two explanations for this warming are attributed to tectonic and magmatic forces. One of the theories is a magmatic super plume inducing a high level of CO_2 into the atmosphere. Carbon dioxide levels in the Cretaceous could have been as high as 3.7 to 14.7 times their present amounts today causing an average 2.8 to 7.7 degrees Celsius. Tectonically, movements of the plates and a sea level fall could cause an additional 4.8 degrees Celsius globally. The combined effect between magmatic and tectonic processes could have placed the Cretaceous Earth 7.6 to 12.5 degrees Celsius higher than today.

A second theory on the warm Cretaceous is the subduction of carbonate materials. By subducting carboniferous materials, a release of carbon dioxide would emit from volcanoes. During the Cretaceous, the Tethys Sea was rich in limestone deposits. By subducting this carboniferous platform, the resulting magma would have become more carbon dioxide rich. Because carbon dioxide dissolves into melts well, it would have remained dissolved until the confining pressure of the magma was low enough to de-gas and release massive quantities of carbon dioxide into the atmosphere causing warming.

Conclusion

Volcanoes represent powerful images and forces on Earth's landscape. Generation of a volcano depends on its location and magmatic origin. Magmas will remain a melt until pressure and temperature allow crystallization and outgassing. During outgassing, the magma chamber will rise and meet Earth's surface causing a volcano. Depending on the composition of the melted material, this volcano could contain a variety of gases. Most of the gases emitted via volcanic eruption are greenhouse gases and cause atmospheric alterations. These atmospheric alterations then force the climate, both regionally and locally, to reach a new equilibrium with the new atmosphere. These changes can reflect as cooling, warming, higher precipitation rates and many others.

Atmospheric Circulation

Atmospheric circulation is the large-scale movement of air, and together with ocean circulation is the means by which thermal energy is redistributed on the surface of the Earth.

The Earth's atmospheric circulation varies from year to year, but the large scale structure of its circulation remains fairly constant. The smaller scale weather systems – mid-latitude depressions, or tropical convective cells – occur "randomly", and long range weather predictions of those cannot be made beyond ten days in practice, or a month in theory. The large scale atmospheric circulation of the Earth, however, is an average of its systems and patterns, and is considered stable over longer periods of time.

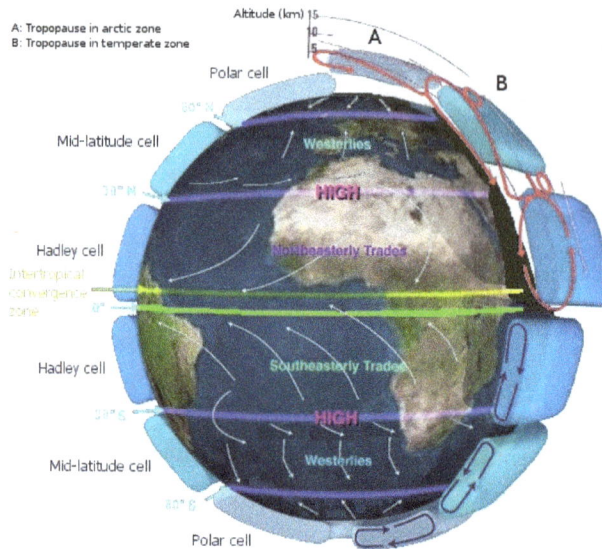

Idealised depiction (at equinox) of large scale atmospheric circulation on Earth.

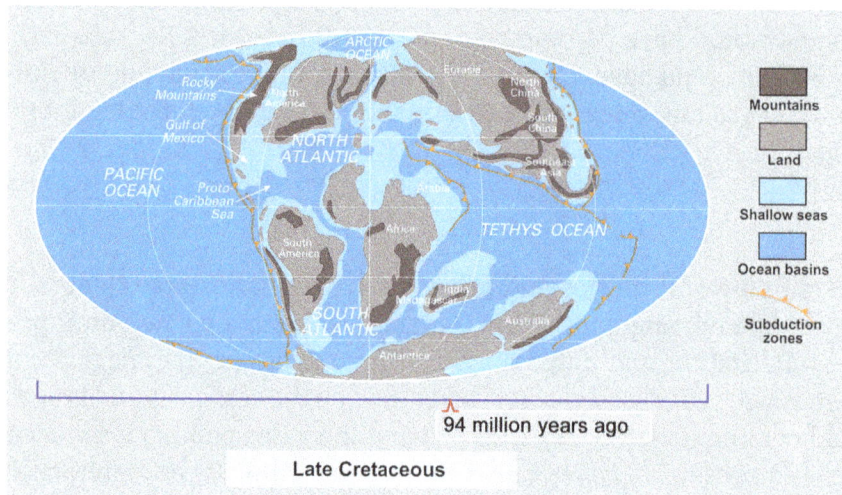

94 million years ago

Late Cretaceous

The Earth's weather is a consequence of its illumination by the Sun, and the laws of thermodynamics. The atmospheric circulation can be viewed, from that standpoint, as a heat engine driven by the Sun's energy, and whose energy sink, ultimately, is the blackness of space. The work produced by that engine causes the motion of the masses of air and in that process it redistributes the energy absorbed by the Earth's surface near the tropics to space and incidentally to the latitudes nearer the poles.

The large scale atmospheric circulation "cells" shift polewards in warmer periods (e.g. interglacials compared to glacials), but remain largely constant as they are, fundamentally, a property of the Earth's size, rotation rate, heating and atmospheric depth, all of which change little. Over very long time periods (hundreds of millions of years), a tectonic uplift can significantly alter their major elements, such as the jet stream, and plate tectonics may shift ocean currents. During the extremely hot climates of the Mesozoic, a third desert belt may have existed at the Equator. But, the overall latitudinal pattern of Earth's climate has not changed.

Latitudinal Circulation Features

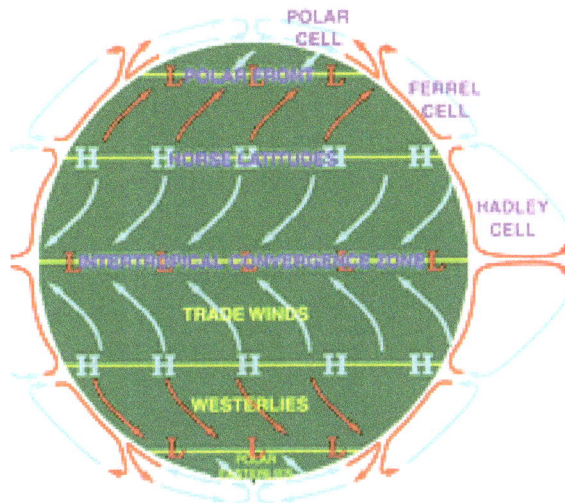

An idealised view of three large circulation cells showing surface winds

The wind belts girdling the planet are organised into three cells; they are from the equator toward the poles, the Hadley cell, the Ferrel cell, and the Polar cell. Those cells exist in both the northern and southern hemispheres. The vast bulk of the atmospheric motion occurs in the Hadley cell. Note that there is one discrete Hadley cell that may split, shift and merge in a complicated process over time. The high pressure systems acting on the Earth's surface is balanced by the low pressure systems elsewhere. As a result, there a balance of forces acting on the Earth's surface.

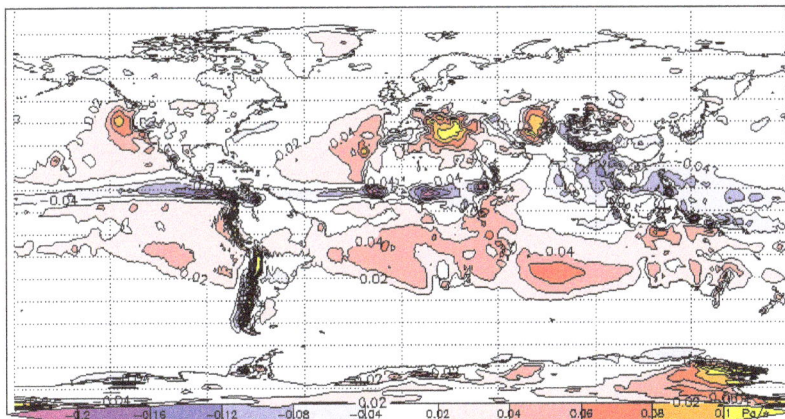

Vertical velocity at 500 hPa, July average. Ascent (negative values) is concentrated close to the solar equator; descent (positive values) is more diffuse but also occurs mainly in the Hadley cell.

Hadley Cell

The atmospheric circulation pattern that George Hadley described was an attempt to explain the trade winds. The Hadley cell is a closed circulation loop, which begins at the equator. There, moist air is warmed by the Earth's surface, decreases in density and rises. A similar air mass rising on the other side of the equator forces those rising air masses to move poleward. The rising air creates a low pressure zone near the equator. As the air moves poleward, it cools, becomes more dense, and descends at about 30th parallel, creating a high-pressure area. The descended air then travels toward the equator along the surface, replacing the air that rose from the equatorial zone, closing the loop of the Hadley cell. The poleward movement of the air in the upper part of the troposphere deviates toward the east due to the coriolis acceleration (a manifestation of conservation of energy). At the ground level however, the movement of the air toward the equator in the lower troposphere deviates toward the west producing a wind from the east (again due to conservation of energy). The winds that flow to the west (from the east, easterly wind) at the ground level in the Hadley cell, are called the Trade Winds.

The ITCZ's band of clouds over the Eastern Pacific and the Americas as seen from space

Though the Hadley cell is described as located at the equator, in the northern hemisphere, it shifts to higher latitudes June and July and toward lower latitudes December and January as it is caused by the Sun's heating of the surface. The zone where the greatest heating takes place is called the "thermal equator". As the southern hemisphere summer is December to March, the movement of the thermal equator to higher southern latitudes takes place then.

The Hadley system provides an example of a thermally direct circulation. The thermodynamic efficiency and power of the Hadley system, considered as a heat engine, is estimated as 2.6% and 200x1012 (TW).

Polar Cell

The Polar cell, likewise, is a simple system. Though cool and dry relative to equatorial air, the air masses at the 60th parallel are still sufficiently warm and moist to undergo convection and drive a thermal loop. At the 60th parallel, the air rises to the tropopause (about 8 km at this latitude) and moves poleward. As it does so, the upper level air mass deviates toward the east. When the air reaches the polar areas, it has cooled and is considerably denser than the underlying air. It descends, creating a cold, dry high-pressure area. At the polar surface level, the mass of air is driven toward the 60th parallel, replacing the air that rose there, and the Polar circulation cell is complete. As the air at the surface moves equatorward, it deviates toward the west. Again, the deviations of the air masses is due to conservation of energy, which is also referred to as the Coriolis effect. The air flows at the surface are called the Polar easterlies (easterly from the east).

The outflow of air mass from the cell creates harmonic waves in the atmosphere known as Rossby waves. These ultra-long waves determine the path of the jet stream, which travels within the transitional zone between the tropopause and the Ferrel cell. By acting as a heat sink, the Polar cell moves the abundant heat from the equator toward the polar regions.

The Hadley cell and the Polar cell are similar in that they are thermally direct; in other words, they exist as a direct consequence of surface temperatures. Their thermal characteristics drive the weather in their domain. The sheer volume of energy that the Hadley cell transports, and the depth of the heat sink that is the Polar cell, ensures that the effects of transient weather phenomena are not only not felt by the system as a whole, but — except under unusual circumstances — do not form. The endless chain of passing highs and lows which is part of everyday life for mid-latitude dwellers, at latitudes between 30 and 60° latitude, is unknown above the 60th and below the 30th parallels. There are some notable exceptions to this rule. In Europe, unstable weather extends to at least 70° north.

These atmospheric features are stable. Even though they may strengthen or weaken regionally over time, they do not vanish entirely.

The Polar cell, orography and Katabatic winds in Antarctica, can create very cold conditions at the surface, for instance the coldest temperature recorded on Earth: −89.2 °C at Vostok Station in Antarctica, measured 1983.

Ferrel Cell

Part of the air rising at 60° latitude diverges at high altitude toward the poles and creates the polar cell. The rest moves toward the equator where it collides at 30° latitude with the high level air of the Hadley cell. There it subsides and strengthens the high pressure ridges beneath. A large part of the energy that drives the Ferrel cell is provided by the Polar and Hadley cells circulating on either side and that drag the Ferrel cell with it. The Ferrel cell, theorized by William Ferrel (1817–1891), is therefore a secondary circulation feature, whose existence depends upon the Hadley and Polar cells on either side of it; it behaves much as an atmospheric ball bearing between the two. It might be thought of as an eddy created by the Hadley and Polar cells. The Ferrel cell is weak and the air flow and temperatures within are variable. For this reason, the mid latitudes are sometimes known as the "zone of mixing." At high altitudes, the Ferrel cell overrides the Hadley and Polar cells. The air of the Ferrel cell that descends at 30° latitude returns poleward at the ground level and as it does so it deviates toward the east. In the upper atmosphere of the Ferrel cell the air moving equatorward, deviates toward the west. Both of those deviations, as in the case of the Hadley and Polar cells, are driven by conservation of energy. As a result, just as the Easterly Trade Winds are found below the Hadley cell, the Westerlies are found beneath the Ferrel cell. The forces driving the flow in the Ferrel cell are weak and so the weather in that zone is variable. Thus, strong high-pressure areas which divert the prevailing westerlies, such as a Siberian high, can override the Ferrel cell, making it discontinuous.

While the Hadley and Polar cells are truly closed loops, the Ferrel cell is not, and the telling point is in the Westerlies, which are more formally known as "the Prevailing Westerlies." The Easterly Trade Winds and the Polar Easterlies have nothing over which to prevail, as their parent circulation cells are strong enough, and face few obstacles either in the form of massive terrain features or high pressure zones. The weaker Westerlies of the Ferrel cell, however, can be disrupted. The local passage of a cold front may change that in a matter of minutes, and frequently does. As a result, at the surface, winds can vary abruptly in direction. But the winds above the surface, where they are less disrupted by terrain, are essentially westerly. A low pressure zone at 60° latitude that moves equator wards, or a high pressure zone at 30°

latitude that moves poleward, will accelerate the Westerlies of the Ferrel cell. A strong high, moving polewards may bring Westerly winds for days. As the high and low pressure zones are established by the Hadley and Polar cells, it can be said that they drive the weather of the middle latitude Ferrel cell.

The Ferrel cell is driven by the Hadley and Polar cells. It has neither a strong source of heat nor a strong cold sink to drive convection. As a result, the weather within the Ferrel cell is highly variable and is influenced by changes to the Hadley and Polar cells. The base of the Ferrel cell is characterized by the movement of air masses, and the location of those air masses is influenced in part by the location of the jet stream, even though it flows near the tropopause. Overall, the movement of surface air is from the 30th latitude to the 60th. However, the upper flow of the Ferrel cell is weak and not well defined.

In contrast to the Hadley and Polar systems, the Ferrel system provides an example of a thermally indirect circulation. The Ferrel system acts as a heat pump with a coefficient of performance of 12.1, consuming kinetic energy at an approximate rate of 275 TW.

Longitudinal Circulation Features

While the Hadley, Ferrel, and Polar cells (whose axes are oriented along parallels i.e. latitudes) are the major features of global heat transport, they do not act alone. Temperature differences also drive a set of circulation cells, whose axis of circulation are longitudinally oriented. This atmospheric motion is known as zonal overturning circulation.

Diurnal Wind Change in Coastal Area

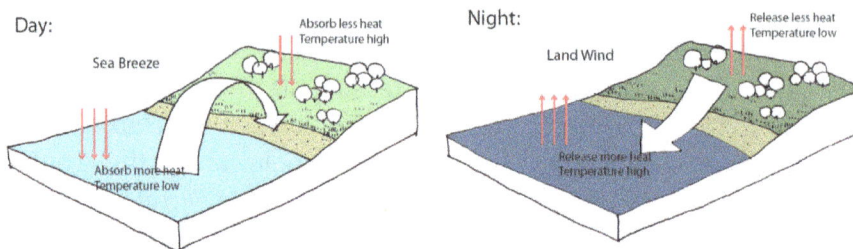

Diurnal wind change in local coastal area, also applies on the continental scale.

Latitudinal circulation is a result of the highest solar radiation per unit area (solar intensity) falling on the tropics. The solar intensity decreases as the latitude increases, reaching essentially zero at the poles. Longitudinal circulation, however, is a result of the heat capacity of water, it's absorptivity, and it's mixing. Water absorbs more heat than does the land, but its temperature does not rise as greatly as does the land. As a result, temperature variations on land are greater than on water. The Hadley, Ferrel, and Polar cells operate at the largest scale of thousands of kilometers (synoptic scale). But, even at mesoscales (a horizontal range of 5 to several hundred kilometres), this effect is noticeable. During the day, air warmed by the relatively hotter land rises, and as it does so it draws a cool breeze from the sea that replaces the risen air. At night, the relatively warmer water and cooler land reverses the process, and a breeze from the land, of air cooled by the land, is carried offshore by night. This described effect is daily (diurnal).

At the larger, synoptic, scale of oceans and continents, this effect is seasonal or even decadal. Warm air rising over the equatorial, continental, and western Pacific Ocean regions. When it reaches the tropopause, it cools and subsides in a region of relatively cooler water mass.

The Pacific Ocean cell plays a particularly important role in Earth's weather. This entirely ocean-based cell comes about as the result of a marked difference in the surface temperatures of the western and eastern Pacific. Under ordinary circumstances, the western Pacific waters are warm and the eastern waters are cool. The process begins when strong convective activity over equatorial East Asia and subsiding cool air off South America's west coast creates a wind pattern which pushes Pacific water westward and piles it up in the western Pacific. (Water levels in the western Pacific are about 60 cm higher than in the eastern Pacific, a difference due entirely to the force of moving air.)

Walker Circulation

The Pacific cell is of such importance that it has been named the Walker circulation after Sir Gilbert Walker, an early-20th-century director of British observatories in India, who sought a means of predicting when the monsoon winds of India would fail. While he was never successful in doing so, his work led him to the discovery of a link between the periodic pressure variations in the Indian Ocean, and those between the eastern and western Pacific, which he termed the "Southern Oscillation".

The movement of air in the Walker circulation affects the loops on either side. Under "normal" circumstances, the weather behaves as expected. But every few years, the winters become unusually warm or unusually cold, or the frequency of hurricanes increases or decreases, and the pattern sets in for an indeterminate period. The Walker Cell plays a key role in this and in the El Niño phenomenon. If convective activity slows in the Western Pacific for some reason (this reason is not currently known), the climate dominoes next to it begin to topple. First, the upper-level westerly winds fail. This cuts off the source of returning, cool air that would normally subside at about 30° north latitude, and therefore the air returning as surface Easterlies ceases.

The consequence of this is twofold. Warm water ceases to surge into the eastern Pacific from the west (it was "piled" by past easterly winds) since there is no longer a surface wind to push it into the area of the west pacific. This and the corresponding effects of the Southern Oscillation result in long-term unseasonable temperatures and precipitation patterns in North and South America, Australia, and Southeast Africa, and the disruption of ocean currents.

Meanwhile, in the Atlantic, fast-blowing upper level Westerlies of the Hadley cell, which would ordinarily be blocked by the Walker circulation and unable to reach such intensities, form. These winds tear apart the tops of nascent hurricanes and greatly diminish the number which are able to reach full strength.

El Niño – Southern Oscillation

El Niño and *La Niña* are opposite surface temperature anomalies of the Southern Pacific, which heavily influence the weather on a large scale. In the case of El Niño, warm surface water approaches the coasts of South America which results in blocking the upwelling of nutrient-rich deep water. This has serious impacts on the fish populations.

In the La Niña case, the convective cell over the western Pacific strengthens inordinately, resulting in colder than normal winters in North America, and a more robust cyclone season in South-East Asia and Eastern Australia. There is also an increased upwelling of deep cold ocean waters and more intense uprising of surface air near South America, resulting in increasing numbers of drought occurrences, although it is often argued that fishermen reap benefits from the more nutrient-filled eastern Pacific waters.

The neutral part of the cycle – the "normal" component – has been referred to humorously by some as "La Nada", which means "the nothing" in Spanish.

Historical Impacts of Climate Change

Climate has affected human life and civilization from the emergence of hominins to the present day. These historical impacts of climate change can improve human life and cause societies to flourish, or can be instrumental in civilization's societal collapse.

Role in Human Evolution

Changes in East African climate have been associated with the evolution of hominins. Researchers have proposed that the regional environment transitioned from humid jungle to more arid grasslands due to tectonic uplift and changes in broader patterns of ocean and atmospheric circulation. This environmental change is believed to have forced hominins to evolve for life in a savannah-type environment. Some data suggest that this environmental change caused the development of modern homimin features; however there exist other data that show that morphological changes in the earliest hominins occurred while the region was still forested. Rapid tectonic uplift likely occurred in the early Pleistocene, changing the local elevation and broadly reorganizing the regional patterns of atmospheric circulation. This can be correlated with the rapid hominin evolution of the Quaternary period. Changes in climate at 2.8, 1.7, and 1.0 million years ago correlate well with observed transitions between recognized hominin species. It is difficult to differentiate correlation from causality in these paleopanthropological and paleoclimatological reconstructions, so these results must be interpreted with caution and related to the appropriate time-scales and uncertainties.

Historic and Prehistoric Societies

The rise and fall of societies have often been linked to environmental factors.

Societal Growth and Urbanization

Approximately one millennium after the 7 ka slowing of sea-level rise, many coastal urban centers rose to prominence around the world. It has been hypothesized that this is correlated with the development of stable coastal environments and ecosystems and an increase in marine productivity (also related to an increase in temperatures), which would provide a food source for hierarchical urban societies.

Societal Collapse

Climate change has been associated with the historical collapse of civilizations, cities and dynasties. Notable examples of this include the Anasazi, Classic Maya, the Harappa, the Hittites, and Ancient Egypt. Other, smaller communities such as the Viking settlement of Greenland have also suffered collapse with climate change being a suggested contributory factor.

The last written records of the Norse Greenlanders are from a 1408 marriage in Hvalsey Church — today the best-preserved of the Norse ruins.

There are two proposed methods of Classic Maya collapse: environmental and non-environmental. The environmental approach uses paleoclimatic evidence to show that movements in the intertropical convergence zone likely caused severe, extended droughts during a few time periods at the end of the archaeological record for the classic Maya. The non-environmental approach suggests that the collapse could be due to increasing class tensions associated with the building of monumental architecture and the corresponding decline of agriculture, increased disease, and increased internal warfare.

The Harappa and Indus civilizations were affected by drought 4,500–3,500 years ago. A decline in rainfall in the Middle East and Northern India 3,800–2,500 is likely to have affected the Hittites and Ancient Egypt.

Historical Era

Notable periods of climate change in recorded history include the Medieval warm period and the little ice age. In the case of the Norse, the Medieval warm period was associated with the Norse age of exploration and arctic colonization, and the later colder periods led to the decline of those colonies.

References

- Royal Society (September 2009). Geoengineering the Climate: Science, Governance and Uncertainty (PDF) (Report). p. 1. ISBN 978-0-85403-773-5. Retrieved 2011-12-01.

- Prentice, I.C.; et al. (2001). "9.2.1 Climate Forcing and Climate Response, in chapter 9. Projections of Future Climate Change". In Houghton J.T.; et al. Climate Change 2001: The Scientific Basis. Contribution of Working Group I to the Third Assessment Report of the Intergovernmental Panel on Climate Change. Cambridge

University Press. ISBN 9780521807678. Retrieved 2010-07-03.

- Schroeder, Daniel V. (2000). An introduction to thermal physics. San Francisco, California: Addison-Wesley. pp. 305–7. ISBN 0-321-27779-1.

- Grosvenor, Edwin S. and Morgan Wesson. Alexander Graham Bell: The Life and Times of the Man Who Invented the Telephone. New York: Harry N. Abrahms, Inc., 1997, p. 274, ISBN 0-8109-4005-1.

- Brian Shmaefsky (2004). Favorite demonstrations for college science: an NSTA Press journals collection. NSTA Press. p. 57. ISBN 978-0-87355-242-4.

- Winfried Henke, Ian Tattersall (eds.); in collaboration with Thorolf Hardt. (2007). Handbook of paleoanthropology. New York: Springer. ISBN 978-3-540-32474-4.

- The Great Warming: Climate Change and the Rise and Fall of Civilizations. New York: Bloomsbury Press. 2008. ISBN 978-1-59691-392-9.

- transl. with introd. by Magnus Magnusson ... (1983). The Vinland sagas: the Norse discovery of America. Harmondsworth, Middlesex: Penguin Books. ISBN 978-0-14-044154-3.

- Hosler D, Sabloff JA, Runge D (1977). "Simulation model development: a case study of the Classic Maya collapse". In Hammond, Norman, Thompson, John L. Social process in Maya prehistory: studies in honour of Sir Eric Thompson. Boston: Academic Press. ISBN 0-12-322050-5.

- Oxford Geoengineering Programme. "Oxford Geoengineering Programme // History of the Oxford Principles". www.geoengineering.ox.ac.uk. Retrieved 2016-02-03.

- "Home | The National Academies of Sciences, Engineering, and Medicine | National-Academies.org | Where the Nation Turns for Independent, Expert Advice". www8.nationalacademies.org. Retrieved 2015-11-24.

- "Climate Intervention Reports » Climate Change at the National Academies of Sciences, Engineering, and Medicine". nas-sites.org. Retrieved 2015-09-02.

- Reynolds, Jesse (2015-08-01). "A critical examination of the climate engineering moral hazard and risk compensation concern". The Anthropocene Review. 2 (2): 174–191. doi:10.1177/2053019614554304. ISSN 2053-0196.

- "Climate Intervention Reports » Climate Change at the National Academies of Sciences, Engineering, and Medicine". nas-sites.org. Retrieved 2015-11-02.

Climate Modelling Analysis

By recording climate patterns over a large period of time, it is possible to create climate models that can predict future climatic patterns such as tidal movements, precipitation and levels of atmospheric pressure. Such models can also help in meteorology and weather sciences. Climatology is best understood in confluence with the major topics listed in the following chapter.

Climate Model

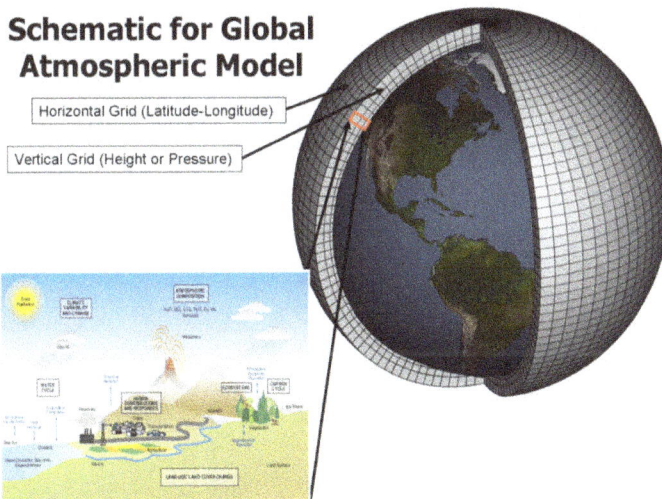

Climate models are systems of differential equations based on the basic laws of physics, fluid motion, and chemistry. To "run" a model, scientists divide the planet into a 3-dimensional grid, apply the basic equations, and evaluate the results. Atmospheric models calculate winds, heat transfer, radiation, relative humidity, and surface hydrology within each grid and evaluate interactions with neighboring points.

Climate models use quantitative methods to simulate the interactions of the important drivers of climate, including atmosphere, oceans, land surface and ice. They are used for a variety of purposes from study of the dynamics of the climate system to projections of future climate.

All climate models take account of incoming energy from the sun as short wave electromagnetic radiation, chiefly visible and short-wave (near) infrared, as well as outgoing long wave (far) infrared electromagnetic. Any imbalance results in a change in temperature.

Models vary in complexity:

- A simple radiant heat transfer model treats the earth as a single point and averages outgoing energy

- This can be expanded vertically (radiative-convective models) and/or horizontally

- Finally, (coupled) atmosphere–ocean–sea ice global climate models solve the full equa-tions for mass and energy transfer and radiant exchange

- Box models can treat flows across and within ocean basins

- Other types of modelling can be interlinked, such as land use, allowing researchers to pre-dict the interaction between climate and ecosystems

Box Models

Box models are simplified versions of complex systems, reducing them to boxes (or reservoirs) linked by fluxes. The boxes are assumed to be mixed homogeneously. Within a given box, the concentration of any chemical species is therefore uniform. However, the abundance of a species within a given box may vary as a function of time due to the input to (or loss from) the box or due to the production, consumption or decay of this species within the box.

Simple box models, i.e. box model with a small number of boxes whose properties (e.g. their volume) do not change with time, are often useful to derive analytical formulas describing the dynamics and steady-state abundance of a species. More complex box models are usually solved using numerical techniques.

Box models are used extensively to model environmental systems or ecosystems and in studies of ocean circulation and the carbon cycle.

Zero-dimensional Models

A very simple model of the radiative equilibrium of the Earth is

$$(1-a)S\pi r^2 = 4\pi r^2 \epsilon \sigma T^4$$

where

- the left hand side represents the incoming energy from the Sun

- the right hand side represents the outgoing energy from the Earth, calculated from the Stefan-Boltzmann law assuming a model-fictive temperature, T, sometimes called the 'equilibrium temperature of the Earth', that is to be found,

and

- S is the solar constant – the incoming solar radiation per unit area—about 1367 $W \cdot m^{-2}$

- a is the Earth's average albedo, measured to be 0.3.

- r is Earth's radius—approximately $6.371 \times 10^6 m$

- π is the mathematical constant (3.141...)

- σ is the Stefan-Boltzmann constant—approximately 5.67×10^{-8} $J \cdot K^{-4} \cdot m^{-2} \cdot s^{-1}$

- ϵ is the effective emissivity of earth, about 0.612

The constant πr^2 can be factored out, giving

$$(1-a)S = 4\epsilon\sigma T^4$$

Solving for the temperature,

$$T = \sqrt[4]{\frac{(1-a)S}{4\epsilon\sigma}}$$

This yields an apparent effective average earth temperature of 288 K (15 °C; 59 °F). This is because the above equation represents the effective *radiative* temperature of the Earth (including the clouds and atmosphere). The use of effective emissivity and albedo account for the greenhouse effect.

This very simple model is quite instructive, and the only model that could fit on a page. For example, it easily determines the effect on average earth temperature of changes in solar constant or change of albedo or effective earth emissivity.

The average emissivity of the earth is readily estimated from available data. The emissivities of terrestrial surfaces are all in the range of 0.96 to 0.99 (except for some small desert areas which may be as low as 0.7). Clouds, however, which cover about half of the earth's surface, have an average emissivity of about 0.5 (which must be reduced by the fourth power of the ratio of cloud absolute temperature to average earth absolute temperature) and an average cloud temperature of about 258 K (−15 °C; 5 °F). Taking all this properly into account results in an effective earth emissivity of about 0.64 (earth average temperature 285 K (12 °C; 53 °F)).

This simple model readily determines the effect of changes in solar output or change of earth albedo or effective earth emissivity on average earth temperature. It says nothing, however about what might cause these things to change. Zero-dimensional models do not address the temperature distribution on the earth or the factors that move energy about the earth.

Radiative-convective Models

The zero-dimensional model above, using the solar constant and given average earth temperature, determines the effective earth emissivity of long wave radiation emitted to space. This can be refined in the vertical to a one-dimensional radiative-convective model, which considers two processes of energy transport:

- upwelling and downwelling radiative transfer through atmospheric layers that both absorb and emit infrared radiation

- upward transport of heat by convection (especially important in the lower troposphere).

The radiative-convective models have advantages over the simple model: they can determine the effects of varying greenhouse gas concentrations on effective emissivity and therefore the surface temperature. But added parameters are needed to determine local emissivity and albedo and address the factors that move energy about the earth.

Effect of ice-albedo feedback on global sensitivity in a one-dimensional radiative-convective climate model.

Higher-simension Models

The zero-dimensional model may be expanded to consider the energy transported horizontally in the atmosphere. This kind of model may well be zonally averaged. This model has the advantage of allowing a rational dependence of local albedo and emissivity on temperature – the poles can be allowed to be icy and the equator warm – but the lack of true dynamics means that horizontal transports have to be specified.

Emics (Earth-system Models of Intermediate Complexity)

Depending on the nature of questions asked and the pertinent time scales, there are, on the one extreme, conceptual, more inductive models, and, on the other extreme, general circulation models operating at the highest spatial and temporal resolution currently feasible. Models of intermediate complexity bridge the gap. One example is the Climber-3 model. Its atmosphere is a 2.5-dimensional statistical-dynamical model with $7.5° \times 22.5°$ resolution and time step of half a day; the ocean is MOM-3 (Modular Ocean Model) with a $3.75° \times 3.75°$ grid and 24 vertical levels.

Gcms (Global Climate Models or General Circulation Models)

General Circulation Models (GCMs) discretise the equations for fluid motion and energy transfer and integrate these over time. Unlike simpler models, GCMs divide the atmosphere and/or oceans into grids of discrete "cells", which represent computational units. Unlike simpler models which make mixing assumptions, processes internal to a cell—such as convection—that occur on scales too small to be resolved directly are parameterised at the cell level, while other functions govern the interface between cells.

Atmospheric GCMs (AGCMs) model the atmosphere and impose sea surface temperatures as boundary conditions. Coupled atmosphere-ocean GCMs (AOGCMs, e.g. HadCM3, EdGCM, GFDL CM2.X, ARPEGE-Climat) combine the two models. The first general circulation climate model that combined both oceanic and atmospheric processes was developed in the late 1960s at the NOAA Geophysical Fluid Dynamics Laboratory AOGCMs represent the pinnacle of complexity in climate models and internalise as many processes as possible. However, they are still under development and uncertainties remain. They may be coupled to models of other processes, such as the carbon cycle, so as to better model feedback effects. Such integrated multi-system models are sometimes referred to as either "earth system models" or "global climate models."

Research and Development

There are three major types of institution where climate models are developed, implemented and used:

- National meteorological services. Most national weather services have a climatology section.

- Universities. Relevant departments include atmospheric sciences, meteorology, climatology, and geography.

- National and international research laboratories. Examples include the National Center for Atmospheric Research (NCAR, in Boulder, Colorado, USA), the Geophysical Fluid Dynamics Laboratory (GFDL, in Princeton, New Jersey, USA), the Hadley Centre for Climate Prediction and Research (in Exeter, UK), the Max Planck Institute for Meteorology in Hamburg, Germany, or the Laboratoire des Sciences du Climat et de l'Environnement (LSCE), France, to name but a few.

The World Climate Research Programme (WCRP), hosted by the World Meteorological Organization (WMO), coordinates research activities on climate modelling worldwide.

A 2012 U.S. National Research Council report discussed how the large and diverse U.S. climate modeling enterprise could evolve to become more unified. Efficiencies could be gained by developing a common software infrastructure shared by all U.S. climate researchers, and holding an annual climate modeling forum, the report found.

Climate Models on the Web

- Dapper/DChart — plot and download model data referenced by the Fourth Assessment Report (AR4) of the Intergovernmental Panel on Climate Change. (No longer available)

- NCAR/UCAR Community Climate System Model (CCSM)

- Do it yourself climate prediction

- Primary research GCM developed by NASA/GISS (Goddard Institute for Space Studies)

- Original NASA/GISS global climate model (GCM) with a user-friendly interface for PCs and Macs

- CCCma model info and interface to retrieve model data

- NOAA/Geophysical Fluid Dynamics Laboratory CM2 global climate model info and model output data files

- University of Victoria Global climate model, free for download. Leading researcher was a contributing author to an IPCC report on climate change.

- vimeo.com/user12523377/videos Visualizations of climate models of ETH Zurich

General Circulation Model

A general circulation model (GCM) is a type of climate model. It employs a mathematical model of the general circulation of a planetary atmosphere or ocean. It uses the Navier–Stokes equations on a rotating sphere with thermodynamic terms for various energy sources (radiation, latent heat).

These equations are the basis for computer programs used to simulate the Earth's atmosphere or oceans. Atmospheric and oceanic GCMs (AGCM and OGCM) are key components along with sea ice and land-surface components.

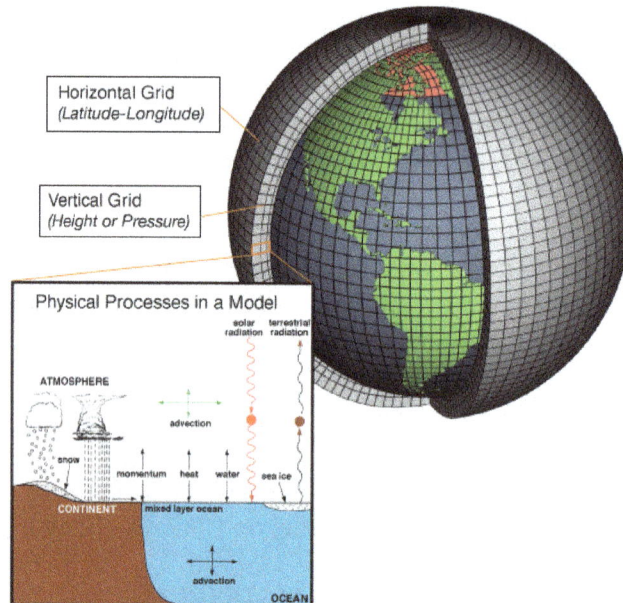

Climate models are systems of differential equations based on the basic laws of physics, fluid motion, and chemistry. To "run" a model, scientists divide the planet into a 3-dimensional grid, apply the basic equations, and evaluate the results. Atmospheric models calculate winds, heat transfer, radiation, relative humidity, and surface hydrology within each grid and evaluate interactions with neighboring points.

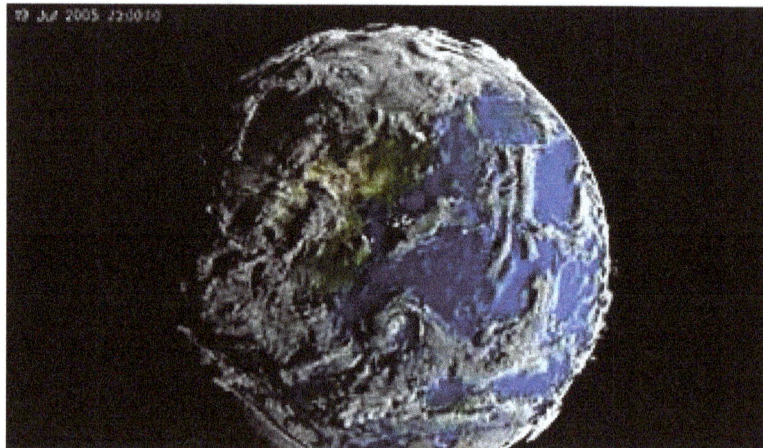

This visualization shows early test renderings of a global computational model of Earth's atmosphere based on data from NASA's Goddard Earth Observing System Model, Version 5 (GEOS-5).

GCMs and global climate models are used for weather forecasting, understanding the climate and forecasting climate change.

Versions designed for decade to century time scale climate applications were originally created by Syukuro Manabe and Kirk Bryan at the Geophysical Fluid Dynamics Laboratory in Princeton, New Jersey. These models are based on the integration of a variety of fluid dynamical, chemical and sometimes biological equations.

Terminology

The acronym *GCM* originally stood for *General Circulation Model*. Recently, a second meaning came into use, namely *Global Climate Model*. While these do not refer to the same thing, General Circulation Models are typically the tools used for modelling climate, and hence the two terms are sometimes used interchangeably. However, the term "global climate model" is ambiguous and may refer to an integrated framework that incorporates multiple components including a general circulation model, or may refer to the general class of climate models that use a variety of means to represent the climate mathematically.

History

In 1956, Norman Phillips developed a mathematical model that could realistically depict monthly and seasonal patterns in the troposphere. It became the first successful climate model. Following Phillips's work, several groups began working to create GCMs. The first to combine both ocean-ic and atmospheric processes was developed in the late 1960s at the NOAA Geophysical Fluid Dynamics Laboratory. By the early 1980s, the United States' National Center for Atmospheric Research had developed the Community Atmosphere Model; this model has been continuously refined. In 1996, efforts began to model soil and vegetation types. Later the Hadley Centre for Climate Prediction and Research's HadCM3 model coupled ocean-atmosphere elements. The role of gravity waves was added in the mid-1980s. Gravity waves are required to simulate regional and global scale circulations accurately.

Atmospheric and Oceanic Models

Atmospheric (AGCMs) and oceanic GCMs (OGCMs) can be coupled to form an atmosphere-ocean coupled general circulation model (CGCM or AOGCM). With the addition of submodels such as a sea ice model or a model for evapotranspiration over land, AOGCMs become the basis for a full climate model.

Trends

A recent trend in GCMs is to apply them as components of Earth system models, e.g. by coupling ice sheet models for the dynamics of the Greenland and Antarctic ice sheets, and one or more chemical transport models (CTMs) for species important to climate. Thus a carbon CTM may allow a GCM to better predict anthropogenic changes in carbon dioxide concentrations. In addition, this approach allows accounting for inter-system feedback: e.g. chemistry-climate models allow the possible effects of climate change on ozone hole to be studied.

Climate prediction uncertainties depend on uncertainties in chemical, physical and social models. Significant uncertainties and unknowns remain, especially regarding the future course of human population, industry and technology.

Structure

Three-dimensional (more properly four-dimensional) GCMs apply discrete equations for fluid motion and integrate these forward in time. They contain parameterisations for processes such as convection that occur on scales too small to be resolved directly.

A simple general circulation model (SGCM) consists of a dynamic core that relates properties such as temperature to others such as pressure and velocity. Examples are programs that solve the primitive equations, given energy input and energy dissipation in the form of scale-dependent friction, so that atmospheric waves with the highest wavenumbers are most attenuated. Such models may be used to study atmospheric processes, but are not suitable for climate projections.

Atmospheric GCMs (AGCMs) model the atmosphere (and typically contain a land-surface model as well) using imposed sea surface temperatures (SSTs). They may include atmospheric chemistry.

AGCMs consist of a dynamical core which integrates the equations of fluid motion, typically for:

- surface pressure
- horizontal components of velocity in layers
- temperature and water vapor in layers
- radiation, split into solar/short wave and terrestrial/infra-red/long wave
- parameters for:
    - convection
    - land surface processes
    - albedo
    - hydrology
    - cloud cover

A GCM contains prognostic equations that are a function of time (typically winds, temperature, moisture, and surface pressure) together with diagnostic equations that are evaluated from them for a specific time period. As an example, pressure at any height can be diagnosed by applying the hydrostatic equation to the predicted surface pressure and the predicted values of temperature between the surface and the height of interest. Pressure is used to compute the pressure gradient force in the time-dependent equation for the winds.

OGCMs model the ocean (with fluxes from the atmosphere imposed) and may contain a sea ice model. For example, the standard resolution of HadOM3 is 1.25 degrees in latitude and longitude, with 20 vertical levels, leading to approximately 1,500,000 variables.

AOGCMs (e.g. HadCM3, GFDL CM2.X) combine the two submodels. They remove the need to specify fluxes across the interface of the ocean surface. These models are the basis for model predictions of future climate, such as are discussed by the IPCC. AOGCMs internalise as many processes as possible. They have been used to provide predictions at a regional scale. While the simpler models are generally susceptible to analysis and their results are easier to understand, AOGCMs may be nearly as hard to analyse as the climate itself.

Grid

The fluid equations for AGCMs are made discrete using either the finite difference method or the spectral method. For finite differences, a grid is imposed on the atmosphere. The simplest grid uses constant angular grid spacing (i.e., a latitude / longitude grid). However, non-rectangular grids (e.g., icosahedral) and grids of variable resolution are more often used. The LMDz model can be arranged to give high resolution over any given section of the planet. HadGEM1 (and other ocean models) use an ocean grid with higher resolution in the tropics to help resolve processes believed to be important for the El Niño Southern Oscillation (ENSO). Spectral models generally use a gaussian grid, because of the mathematics of transformation between spectral and grid-point space. Typical AGCM resolutions are between 1 and 5 degrees in latitude or longitude: HadCM3, for example, uses 3.75 in longitude and 2.5 degrees in latitude, giving a grid of 96 by 73 points (96 x 72 for some variables); and has 19 vertical levels. This results in approximately 500,000 "basic" variables, since each grid point has four variables (u,v, T, Q), though a full count would give more (clouds; soil levels). HadGEM1 uses a grid of 1.875 degrees in longitude and 1.25 in latitude in the atmosphere; HiGEM, a high-resolution variant, uses 1.25 x 0.83 degrees respectively. These resolutions are lower than is typically used for weather forecasting. Ocean resolutions tend to be higher, for example HadCM3 has 6 ocean grid points per atmospheric grid point in the horizontal.

For a standard finite difference model, uniform gridlines converge towards the poles. This would lead to computational instabilities and so the model variables must be filtered along lines of latitude close to the poles. Ocean models suffer from this problem too, unless a rotated grid is used in which the North Pole is shifted onto a nearby landmass. Spectral models do not suffer from this problem. Some experiments use geodesic grids and icosahedral grids, which (being more uniform) do not have pole-problems. Another approach to solving the grid spacing problem is to deform a Cartesian cube such that it covers the surface of a sphere.

Flux Buffering

Some early versions of AOGCMs required an *ad hoc* process of "flux correction" to achieve a stable climate. This resulted from separately prepared ocean and atmospheric models that each used an implicit flux from the other component different than that component could produce. Such a model failed to match observations. However, if the fluxes were 'corrected', the factors that led to these unrealistic fluxes might be unrecognised, which could affect model sensitivity. As a result, the vast majority of models used in the current round of IPCC reports do not use them. The model improvements that now make flux corrections unnecessary include improved ocean physics, improved resolution in both atmosphere and ocean, and more physically consistent coupling between atmosphere and ocean submodels. Improved models now maintain stable, multi-century simulations of surface climate that are considered to be of sufficient quality to allow their use for climate projections.

Convection

Moist convection releases latent heat and is important to the Earth's energy budget. Convection occurs on too small a scale to be resolved by climate models, and hence it must be handled via parameters. This has been done since the 1950s. Akio Arakawa did much of the early work, and variants of his scheme are still used, although a variety of different schemes are now in use. Clouds are also typically handled with a parameter, for a similar lack of scale. Limited understanding of clouds has limited the success of this strategy, but not due to some inherent shortcoming of the method.

Software

Most models include software to diagnose a wide range of variables for comparison with observations or study of atmospheric processes. An example is the 1.5-metre temperature, which is the standard height for near-surface observations of air temperature. This temperature is not directly predicted from the model but is deduced from surface and lowest-model-layer temperatures. Other software is used for creating plots and animations.

Projections

Projected annual mean surface air temperature from 1970-2100, based on SRES emissions scenario A1B, using the NOAA GFDL CM2.1 climate model (credit: NOAA Geophysical Fluid Dynamics Laboratory).

Coupled AOGCMs use transient climate simulations to project/predict climate changes under various scenarios. These can be idealised scenarios (most commonly, CO_2 emissions increasing at 1%/yr) or based on recent history (usually the "IS92a" or more recently the SRES scenarios). Which scenarios are most realistic remains uncertain.

The 2001 IPCC Third Assessment Report shows the global mean response of 19 different coupled models to an idealised experiment in which emissions increased at 1% per year. It shows the response of a smaller number of models to more recent trends. For the 7 climate models shown there, the temperature change to 2100 varies from 2 to 4.5 °C with a median of about 3 °C.

Future scenarios do not include unknown events – for example, volcanic eruptions or changes in solar forcing. These effects are believed to be small in comparison to greenhouse gas (GHG) forcing in the long term, but large volcanic eruptions, for example, can exert a substantial temporary cooling effect.

Human GHG emissions are a model input, although it is possible to include an economic/technological submodel to provide these as well. Atmospheric GHG levels are usually supplied as an input, though it is possible to include a carbon cycle model that reflects vegetation and oceanic processes to calculate such levels.

Emissions Scenarios

Projected change in annual mean surface air temperature from the late 20th century to the middle 21st century, based on SRES emissions scenario A1B (credit: NOAA Geophysical Fluid Dynamics Laboratory).

For the six SRES marker scenarios, IPCC (2007) gave a "best estimate" of global mean temperature increase (2090–2099 relative to the period 1980–1999) of 1.8 °C to 4.0 °C. Over the same time period, the "likely" range (greater than 66% probability, based on expert judgement) for these scenarios was for a global mean temperature increase of 1.1 to 6.4 °C.

In 2008 a study made climate projections using several emission scenarios. In a scenario where global emissions start to decrease by 2010 and then declined at a sustained rate of 3% per year, the likely global average temperature increase was predicted to be 1.7 °C above pre-industrial levels by 2050, rising to around 2 °C by 2100. In a projection designed to simulate a future where no efforts are made to reduce global emissions, the likely rise in global average temperature was predicted to be 5.5 °C by 2100. A rise as high as 7 °C was thought possible, although less likely.

Another no-reduction scenario resulted in a median warming over land (2090–99 relative to the period 1980–99) of 5.1 °C. Under the same emissions scenario but with a different model, the predicted median warming was 4.1 °C.

Model Accuracy

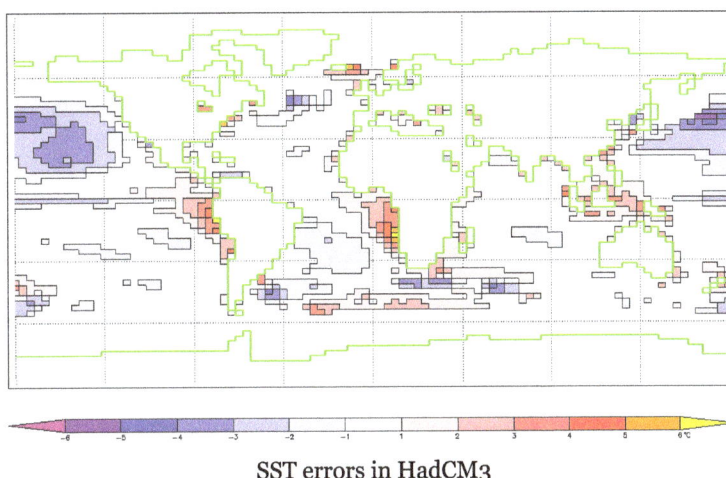

SST errors in HadCM3

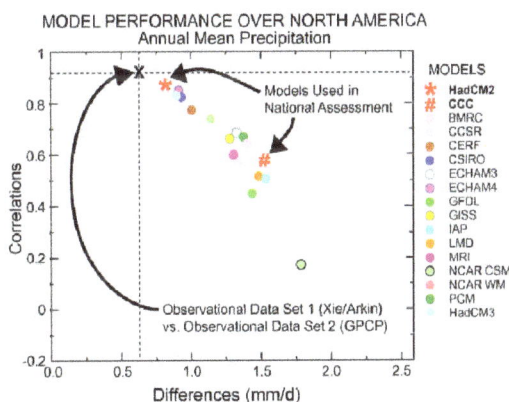

North American precipitation from various models.

AOGCMs internalise as many processes as are sufficiently understood. However, they are still under development and significant uncertainties remain. They may be coupled to models of other processes, such as the carbon cycle, so as to better model feedbacks. Most recent simulations show "plausible" agreement with the measured temperature anomalies over the past 150 years, when driven by observed changes in greenhouse gases and aerosols. Agreement improves by including both natural and anthropogenic forcings.

Imperfect models may nevertheless produce useful results. GCMs are capable of reproducing the general features of the observed global temperature over the past century.

A debate over how to reconcile climate model predictions that upper air (tropospheric) warming should be greater than observed surface warming, some of which appeared to show otherwise, was resolved in favour of the models, following data revisions.

Global Warming Projections

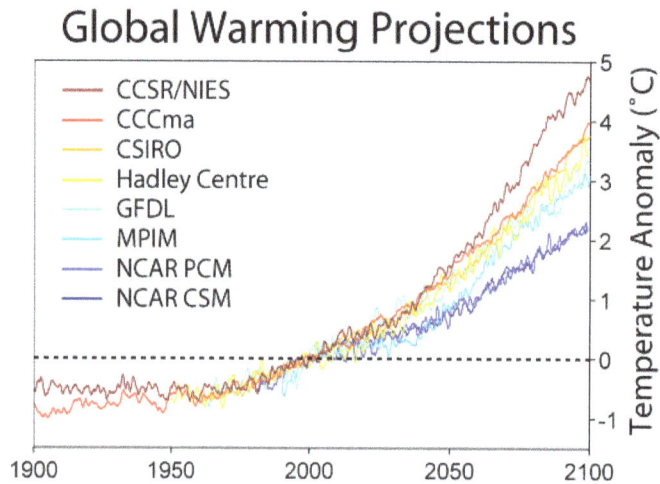

Temperature predictions from some climate models assuming the SRES A2 emissions scenario.

Cloud effects are a significant area of uncertainty in climate models. Clouds have competing effects on climate. They cool the surface by reflecting sunlight into space; they warm it by increasing the amount of infrared radiation transmitted from the atmosphere to the surface. In the 2001 IPCC report possible changes in cloud cover were highlighted as a major uncertainty in predicting climate.

Climate researchers around the world use climate models to understand the climate system. Thousands of papers have been published about model-based studies. Part of this research is to improve the models.

In 2000, a comparison between measurements and dozens of GCM simulations of ENSO-driven tropical precipitation, water vapor, temperature, and outgoing longwave radiation found similarity between measurements and simulation of most factors. However the simulated change in precipitation was about one-fourth less than what was observed. Errors in simulated precipitation imply errors in other processes, such as errors in the evaporation rate that provides moisture to create precipitation. The other possibility is that the satellite-based measurements are in error. Either indicates progress is required in order to monitor and predict such changes.

A more complete discussion of climate models is provided in the IPCC's Third Assessment Report.

- The model mean exhibits good agreement with observations.

- The individual models often exhibit worse agreement with observations.

- Many of the non-flux adjusted models suffered from unrealistic climate drift up to about 1 °C/century in global mean surface temperature.

- The errors in model-mean surface air temperature rarely exceed 1 °C over the oceans and 5 °C over the continents; precipitation and sea level pressure errors are relatively greater but the magnitudes and patterns of these quantities are recognisably similar to observations.

- Surface air temperature is particularly well simulated, with nearly all models closely matching the observed magnitude of variance and exhibiting a correlation > 0.95 with the observations.

- Simulated variance of sea level pressure and precipitation is within ±25% of observed.

- All models have shortcomings in their simulations of the present day climate of the stratosphere, which might limit the accuracy of predictions of future climate change.

  o There is a tendency for the models to show a global mean cold bias at all levels.

  o There is a large scatter in the tropical temperatures.

  o The polar night jets in most models are inclined poleward with height, in noticeable contrast to an equatorward inclination of the observed jet.

  o There is a differing degree of separation in the models between the winter sub-tropical jet and the polar night jet.

- For nearly all models the r.m.s. error in zonal- and annual-mean surface air temperature is small compared with its natural variability.

  o There are problems in simulating natural seasonal variability. (2000)

    ▪ In flux-adjusted models, seasonal variations are simulated to within 2 K of observed values over the oceans. The corresponding average over non-flux-adjusted models shows errors up to about 6 K in extensive ocean areas.

    ▪ Near-surface land temperature errors are substantial in the average over flux-adjusted models, which systematically underestimates (by about 5 K) temperature in areas of elevated terrain. The corresponding average over non-flux-adjusted models forms a similar error pattern (with somewhat increased amplitude) over land.

    ▪ In Southern Ocean mid-latitudes, the non-flux-adjusted models overestimate the magnitude of January-minus-July temperature differences by ~5 K due to an overestimate of summer (January) near-surface temperature. This error is common to five of the eight non-flux-adjusted models.

    ▪ Over Northern Hemisphere mid-latitude land areas, zonal mean differences between July and January temperatures simulated by the non-flux-adjusted models show a greater spread (positive and negative) about observed values than results from the flux-adjusted models.

    ▪ The ability of coupled GCMs to simulate a reasonable seasonal cycle is a necessary condition for confidence in their prediction of long-term climatic changes (such as global warming), but it is not a sufficient condition unless the seasonal cycle and long-term changes involve similar climatic processes.

- Coupled climate models do not simulate with reasonable accuracy clouds and some related hydrological processes (in particular those involving upper tropospheric humidity). Problems in the simulation of clouds and upper tropospheric humidity, remain worrisome because the associated processes account for most of the uncertainty in climate model simulations of anthropogenic change.

The precise magnitude of future changes in climate is still uncertain; for the end of the 21st century (2071 to 2100), for SRES scenario A2, the change of global average SAT change from AOGCMs compared with 1961 to 1990 is +3.0 °C (5.4 °F) and the range is +1.3 to +4.5 °C (+2.3 to 8.1 °F).

The IPCC's Fifth Assessment Report asserted "...very high confidence that models reproduce the general features of the global-scale annual mean surface temperature increase over the historical period." However, the report also observed that the rate of warming over the period 1998-2012 was lower than that predicted by 111 out of 114 Coupled Model Intercomparison Project climate models.

Relation to Weather Forecasting

The global climate models used for climate projections are similar in structure to (and often share computer code with) numerical models for weather prediction, but are nonetheless logically distinct.

Most weather forecasting is done on the basis of interpreting numerical model results. Since forecasts are short—typically a few days or a week—such models do not usually contain an ocean model but rely on imposed SSTs. They also require accurate initial conditions to begin the forecast—typically these are taken from the output of a previous forecast, blended with observations. Predictions must require only a few hours; but because they only cover a one-week the models can be run at higher resolution than in climate mode. Currently the ECMWF runs at 40 km (25 mi) resolution as opposed to the 100-to-200 km (62-to-124 mi) scale used by typical climate model runs. Often local models are run using global model results for boundary conditions, to achieve higher local resolution: for example, the Met Office runs a mesoscale model with an 11 km (6.8 mi) resolution covering the UK, and various agencies in the US employ models such as the NGM and NAM models. Like most global numerical weather prediction models such as the GFS, global climate models are often spectral models instead of grid models. Spectral models are often used for global models because some computations in modeling can be performed faster, thus reducing run times.

Computations

Climate models use quantitative methods to simulate the interactions of the atmosphere, oceans, land surface and ice.

All climate models take account of incoming energy as short wave electromagnetic radiation, chiefly visible and short-wave (near) infrared, as well as outgoing energy as long wave (far) infrared electromagnetic radiation from the earth. Any imbalance results in a change in temperature.

The most talked-about models of recent years relate temperature to emissions of greenhouse gases. These models project an upward trend in the surface temperature record, as well as a more rapid increase in temperature at higher altitudes.

Three (or more properly, four since time is also considered) dimensional GCM's discretise the equations for fluid motion and energy transfer and integrate these over time. They also contain parametrisations for processes such as convection that occur on scales too small to be resolved directly.

Atmospheric GCMs (AGCMs) model the atmosphere and impose sea surface temperatures as boundary conditions. Coupled atmosphere-ocean GCMs (AOGCMs, e.g. HadCM3, EdGCM, GFDL CM2.X, ARPEGE-Climat) combine the two models.

Models range in complexity:

- A simple radiant heat transfer model treats the earth as a single point and averages outgoing energy

- This can be expanded vertically (radiative-convective models), or horizontally

- Finally, (coupled) atmosphere–ocean–sea ice global climate models discretise and solve the full equations for mass and energy transfer and radiant exchange.

- Box models treat flows across and within ocean basins.

Other submodels can be interlinked, such as land use, allowing researchers to predict the interaction between climate and ecosystems.

Other Climate Models

Earth-system Models of Intermediate Complexity (Emics)

The Climber-3 model uses a 2.5-dimensional statistical-dynamical model with 7.5° × 22.5° resolution and time step of 1/2 a day. An oceanic submodel is MOM-3 (Modular Ocean Model) with a 3.75° × 3.75° grid and 24 vertical levels.

Radiative-convective Models (RCM)

One-dimensional, radiative-convective models were used to verify basic climate assumptions in the '80s and '90s.

Mesothermal

In climatology, the term mesothermal is used to refer to certain forms of climate found typically in the Earth's Temperate Zones. It has a moderate amount of heat, with winters not cold enough to sustain snow cover. Summers are warm within oceanic climate regimes, and hot within continental climate regimes.

Origin of Term

The term is derived from two Greek words meaning "having a moderate amount of heat." This can be misinterpreted, however, since the term is actually intended to describe only the temperature conditions that prevail during the winter months, rather than those for the year as a whole.

Definition

Under the broadest definition, all places with an average temperature in their coldest month that is colder than 18°C, but warmer than −3°C, are said to have a mesothermal climate. In some climate classification schemes, however, this is divided into two segments, with a coldest-month average of 6°C being the line of demarcation between them; then only those locations with a coldest-month temperature of between −3°C and 6°C are reckoned as mesothermal, the label "subtropical" being applied to areas where the average temperature in the coldest month ranges from 6°C to 18°C.

Observing the narrower definition articulated above, the mesothermal locations are those where the winters are too cold to allow year-round photosynthesis, but not cold enough to support a fixed period of continuous snow cover every year.

Range

In the USA, the northern boundary line between mesothermal and microthermal ranges is north of Juneau and Sitka at the Pacific Ocean. It goes sharply south to about 38N latitude in the Rockies, then eastward across the lower Midwest to the East Coast near Boston. The southern boundary line between mesothermal and megathermal (or tropical) is across south Florida just above Palm Beach.

Summer

Summers in these places may be hot (that is to say, having an average temperature in their warmest month of 22°C or above) or merely warm (with the warmest month averaging between 10°C and 22°C). The hot-summer, or continental, mesothermal climate is encountered exclusively in the Northern Hemisphere, in the landmass interiors of Asia and North America and along their east coasts, while the most frequently seen example of a warm to cool-summer mesothermal climate is the oceanic climates found along the west coasts of all of the world's continents, roughly equidistant between the geographical tropical and polar zones.

List of mesothermal cities and their summer temperatures:
Hong Kong-hot summer
Milan-hot summer
New York City-hot summer
Tokyo-hot summer
London-warm summer
Mexico City-warm summer
Vancouver-warm summer

Moisture

In addition to being subdivisible by summer temperature, mesothermal climates can also be sub-classified on the basis of precipitation — into humid, semiarid and arid subtypes within your front door within depth

Microthermal

The word *microthermal* is derived from two Greek words meaning "small" and "heat". This is misleading, however, since the term is intended to describe only the temperature conditions that prevail during the winter months, rather than those of the entire year.

The characteristic feature of the microthermal climate is cold winters — specifically, winters that are cold enough to ensure that snow will remain on the ground continuously for a fixed period of time every year. Conceptually, an average temperature of 0°C or colder is assumed to be necessary to bring this about; thus the climate of a location where at least one full month is this cold is classified as microthermal (however, at least one month in the summer must average 10°C or higher; otherwise the climate would be reckoned as polar). This definition places all of the world's microthermal climates in the Northern Hemisphere, as the absence of broad land masses at upper-middle latitudes in the Southern Hemisphere precludes the existence of such temperature conditions there.

Microthermal climates are typically subdivided into three categories based on the temperature characteristics of the summer season. The southernmost of the three is frequently referred to as the temperate continental climate, and has hot summers — that is to say, at least one month has an average temperature of 22°C (71.6°F) or above. The middle zone is often labelled hemiboreal, and no summer month there has an average temperature as warm as 22°C, but at least four months will average 10°C (50°F) or higher. The northernmost of the three microthermal zones is the subarctic, or boreal zone; there only one to three months will have average temperatures of at least 10°C.

In North America, microthermal climates commence north of Boston along the Atlantic seaboard, this line drifting gradually southward further inland, reaching approximately 38° at the eastern edge of the Rocky Mountains, then curving dramatically northward near the Pacific coast, reaching the Pacific Ocean just south of Juneau, Alaska. In Asia, the latitude at which these climates begin is several degrees lower due to the pervasive influence of the vast Siberian anticyclone, or high-pressure system, and in continental Europe the line actually runs longitudinally rather than latitudinally, cutting through central Poland after beginning north of the Arctic Circle along the Norwegian coast, thereafter moving diagonally across Scandinavia.

The boundary between the microthermal and polar climate zones is farthest north in western Europe (actually within the Arctic Circle there), and farthest south along the east coast of North America (at about 56° North latitude on the central coast of Labrador); it then trends northward across Canada before dropping south again as it courses through Alaska. Throughout most of Siberia, the boundary tends to follow the Arctic Circle fairly closely.

In addition to having various summer temperature regimes, microthermal climates also differ from one another in how much precipitation they receive — such climates may be humid, semiarid or arid. Most of the Turkestan-Gobi desert system has an arid microthermal climate, while the best-known example of the semiarid microthermal climate can be found in the "steppes of Central Asia" immortalized by Russian classical music composer Alexander Borodin.

Megathermal

In climatology, the term megathermal (or less commonly, macrothermal) is sometimes used as a synonym for "tropical."

In order for a particular place to qualify as having a megathermal climate, every single month out of the year must have an average temperature of 18°C or above.

Megathermal climates are sometimes split into two temperature-based subsets — equatorial and tropical (the latter used here in the sense of "outer tropical") — with "equatorial" denoting little or no variation in temperature throughout the year and "tropical" denoting significant seasonal variation, even though no month has an average temperature of below 18°C. In addition, what temperature fluctuations do exist in an equatorial climate will typically bear no relationship to the astronomical seasons for the applicable side of the equator, while in the (outer) tropical subtype the temperature will move in concert with the seasons, the time of higher sun and longer days being warmest and the time of lower sun and shorter days coolest.

These climates can also be subdivided on the basis of rainfall, as examples of humid, semiarid and arid places can all be found within the megathermal category (although the region of arid megathermal climate is small compared with the extent of deserts existing in other climate zones, particularly the neighboring subtropical zone).

References

- Lynch, Peter (2006). "The ENIAC Integrations". The Emergence of Numerical Weather Prediction. Cambridge University Press. pp. 206–208. ISBN 978-0-521-85729-1.

- "Pubs.GISS: Sun and Hansen 2003: Climate simulations for 1951-2050 with a coupled atmosphere-ocean model". pubs.giss.nasa.gov. 2003. Retrieved 2015-08-25.

- Xue, Yongkang & Michael J. Fennessey (20 March 1996). "Impact of vegetation properties on U.S. summer weather prediction" (PDF). Journal of Geophysical Research. American Geophysical Union. 101 (D3): 7419. Bibcode:1996JGR...101.7419X. doi:10.1029/95JD02169. Retrieved 6 January 2011.

- Ken, Richard A (13 April 2001). "Global Warming: Rising Global Temperature, Rising Uncertainty". Science. 292 (5515): 192–194. doi:10.1126/science.292.5515.192. PMID 11305301. Retrieved 20 April 2010.

- "Atmospheric Model Intercomparison Project". The Program for Climate Model Diagnosis and Intercomparison, Lawrence Livermore National Laboratory. Retrieved 21 April 2010.

- "High Resolution Global Environmental Modelling (HiGEM) home page". Natural Environment Research Council and Met Office. 18 May 2004. Retrieved 5 October 2010.

- C. Jablonowski , M. Herzog , J. E. Penner , R. C. Oehmke , Q. F. Stout , B. van Leer, January 1960.5091 "Adaptive Grids for Weather and Climate Models" (2004).

Climate Pattern and Climatology Research

Climate changes along periods that recur over and over and these repetitive "oscillatory" patterns influence the habitation and growth of vegetation, animals and human beings. This chapter concentrates on climate patterns found near the different continents as well as on classifications such as climate state that has been in use in climatology studies.

Climate Pattern

A climate pattern may come in the form of a regular cycle, like the diurnal cycle or the seasonal cycle; a quasi periodic event, like El Niño; or a highly irregular event, such as a volcanic winter. The regular cycles are generally well understood and may be removed by normalization. For example, graphs which show trends of temperature change will usually have the effects of seasonal variation removed.

Modes of Variability

A *mode of variability* is a climate pattern with identifiable characteristics, specific regional effects, and often oscillatory behavior. Many modes of variability are used by climatologists as indices to represent the general climatic state of a region affected by a given climate pattern.

Measured via an empirical orthogonal function analysis, the mode of variability with the greatest effect on climates worldwide is the seasonal cycle, followed by El Niño-Southern Oscillation, followed by thermohaline circulation.

Other well-known modes of variability include:

- The Antarctic oscillation
- The Arctic oscillation
- The Atlantic multidecadal oscillation
- The Indian Ocean Dipole
- The Madden–Julian oscillation
- The North Atlantic oscillation
- The Pacific decadal oscillation
- The Pacific-North American teleconnection pattern
- The Quasi-biennial oscillation

Climate Oscillation

A climate oscillation or climate cycle is any recurring cyclical oscillation within global or regional climate, and is a type of climate pattern. These fluctuations in atmospheric temperature, sea surface temperature, precipitation or other parameters can be quasi-periodic, often occurring on inter-annual, multi-annual, decadal, multidecadal, century-wide, millennial or longer timescales. They are not perfectly periodic and a Fourier analysis of the data does not give a sharp spectrum.

A prominent example is the El Niño Southern Oscillation, involving sea surface temperatures along a stretch of the equatorial Central and East Pacific Ocean and the western coast of tropical South America, but which affects climate worldwide.

Records of past climate conditions are recovered through geological examination of proxies, found in glacier ice, sea bed sediment, tree ring studies or otherwise.

Examples

Many oscillations on different time-scales are hypothesized, although the causes may be unknown. (Some of them are more like a random walk than an oscillation.) Here is a list of known or proposed climatic oscillations:

- the glacial periods of the last ice age – period around 100 000 years
- North African climate cycles – tens of thousands of years
- the Atlantic Multidecadal Oscillation – around 50 to 70 years, but unpredictable
- the El Niño Southern Oscillation – 2 to 7 years
- the Pacific decadal oscillation – 8 to 12 years? (not clear)
- the Interdecadal Pacific Oscillation – 15 to 30 years? (not clear)
- the Arctic oscillation – no particular periodicity
- the North Atlantic Oscillation – no particular periodicity
- the North Pacific Oscillation – ?
- the Hale cycle or sunspot cycle – about 11 years
- the Quasi-biennial oscillation – about 30 months
- a 60-year climate cycle recorded in many ancient calendars

Anomalies in oscillations sometimes occur when they coincide, as in the Arctic dipole anomaly (a combination of the Arctic and North Atlantic oscillations) and the longer-term Younger Dryas, a sudden non-linear cooling event that occurred at the onset of the current Holocene interglacial.

In the case of volcanoes, large eruptions such as Mount Tambora in 1816, which led to the Year Without a Summer, typically cool the climate, especially when the volcano is located in the tropics. Around 70 000 years ago the Toba supervolcano eruption created an especially cold period during the ice age, leading to a possible genetic bottleneck in human populations. However, outgassing from large igneous provinces such as the Permian Siberian Traps can input carbon dioxide into the atmosphere, warming the climate. Triggering of other mechanisms, such as methane clathrate deposits as during the Paleocene-Eocene Thermal Maximum, increased the rate of climatic temperature change and oceanic extinctions.

Another longer-term near-millennial oscillation involves the Daansgard-Oeschger cycles, occurring on roughly 1,500-year cycles during the last glacial maximum. They may be related to the Holocene Bond events, and may involve factors similar to those responsible for Heinrich events.

Origins and Causes

There are close correlations between Earth's climate oscillations and astronomical factors (barycenter changes, solar variation, cosmic ray flux, cloud albedo feedback, Milankovic cycles), and modes of heat distribution between the ocean-atmosphere climate system. In some cases, current, historical and paleoclimatological natural oscillations may be masked by significant volcanic eruptions, impact events, irregularities in climate proxy data, positive feedback processes or anthropogenic emissions of substances such as greenhouse gases.

Effects

Extreme phases of short-term climate oscillations such as ENSO can result in characteristic patterns of floods and droughts (including megadroughts), monsoonal disruption and extreme temperatures in the form of heat waves and cold waves. Shorter-term climate oscillations typically do not directly result in longer-term climate change in temperatures. However, the effects of underlying climate trends such as recent global warming and oscillations can be cumulative to global temperature, producing shorter-term fluctuations in the instrumental and satellite temperature records.

Collapses of past civilizations such as the Maya may be related to cycles of precipitation, especially drought, that in this example also correlates to the Western Hemisphere Warm Pool.

One example of possible correlations between factors affecting the climate and global events, popular with the media, is a 2003 study on the correlation between wheat prices and sunspot numbers.

Analysis and Uncertainties

Radiative forcings and other factors in a climate oscillation must obey the laws of atmospheric thermodynamics. However, because Earth's climate is inherently a complex system, simple Fourier analysis or climate modelling often does not create a perfect replication of the observed or inferred conditions. No climate cycle is found to be perfectly periodic, although the Milankovich cycles (based on multiple superimposed orbital cycles and Earth's precession) are quite close to being periodic (perhaps almost periodic?).

One difficulty in detecting climate cycles is that the Earth's climate has been changing in non-cyclic ways over most paleoclimatological timescales. For instance, we are now in a period of global warming that appears anthropogenic. In a larger timeframe, the Earth is emerging from the latest ice age, cooling from the Holocene climatic optimum and warming from the so-called "Little Ice Age", which means that climate has been constantly changing over the last 15,000 years or so. During warm periods, temperature fluctuations are often of a lesser amplitude. The Pleistocene period, dominated by repeated glaciations, developed out of more stable conditions in the Miocene and Pliocene climate. Holocene climate has been relatively stable. All of these changes complicate the task of looking for cyclical behavior in the climate.

Positive feedback, negative feedback, and ecological inertia from the land-ocean-atmosphere system often attenuate or reverse smaller effects, whether from orbital forcings, solar variations or changes in concentrations of greenhouse gases. Most climatologists recognize the existence of various tipping points that push small forcings beyond a certain threshold that makes the change irreversible while the forcings are still in place. Certain feedbacks involvong processes such as clouds are also uncertain; for contrails, natural cirrus clouds, oceanic dimethyl sulfide and a land-based equivalent, competing theories exist concerning effects on climatic temperatures, for example contrasting the Iris hypothesis and CLAW hypothesis.

Through Geologic and Historical Time

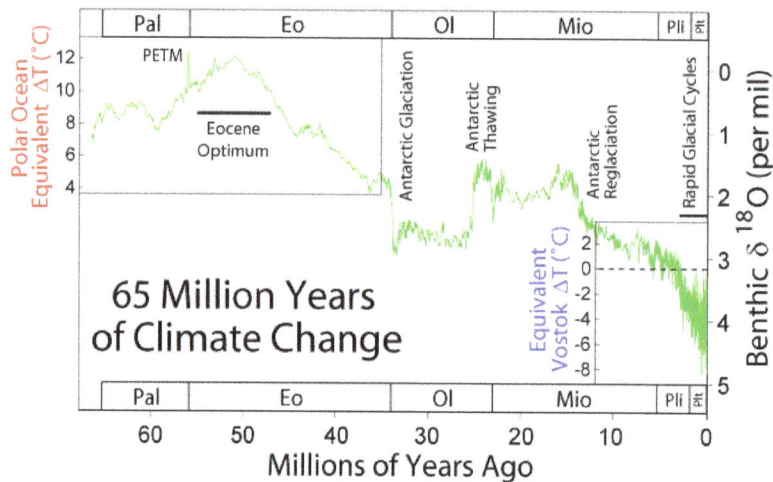

Climate change over the past 65 million years, using proxy data including Oxygen-18 ratios from foraminifera.

Various climate forcings are typically in flux throughout geologic time, and some processes of the Earth's temperature may be self-regulating. For example, during the Snowball Earth period, large glacial ice sheets spanned to Earth's equator, covering nearly its entire surface, and very low albedo created extremely low temperatures, while the accumulation of snow and ice likely removed carbon dioxide through atmospheric deposition. However, the absence of plant cover to absorb atmospheric CO_2 emitted by volcanoes meant that the greenhouse gas could accumulate in the atmosphere. There was also an absence of exposed silicate rocks, which use CO_2 when they undergo weathering. This created a warming that later melted the ice and brought Earth's temperature back to equilibrium. During the following eons of the Paleozoic, cosmic ray flux and occasional nearby supernova explosions (one hypothesis for the cause of the Ordovician–Silurian extinction event) and gamma ray bursts may have induced ice ages or other sudden climate changes.

Holocene Temperature Variations

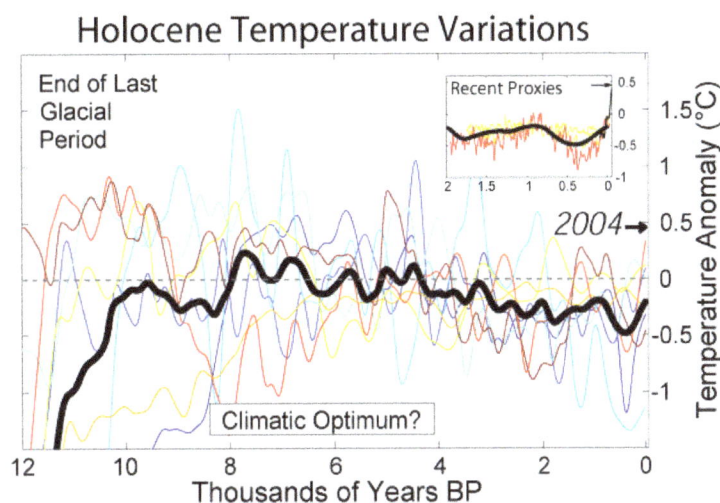

Temperature change over the past 12 000 years, from various sources. The thick black curve is an average.

Throughout the Cenozoic, multiple climate forcings led to warming and cooling of the atmosphere, which led to the early formation of the Antarctic ice sheet, subsequent melting, and its later reglaciation. The temperature changes occurred somewhat suddenly, at carbon dioxide concentrations of about 600–760 ppm and temperatures approximately 4 °C warmer than today. During the Pleistocene, cycles of glaciations and interglacials occurred on cycles of roughly 100,000 years, but may stay longer within an interglacial when orbital eccentricity approaches zero, as during the current interglacial. Previous interglacials such as the Eemian phase created temperatures higher than today, higher sea levels, and some partial melting of the West Antarctic ice sheet. The warmest part of the current interglacial occurred during the early Holocene Optimum, when temperatures were a few degrees Celsius warmer than today, and a strong African Monsoon created grassland conditions in the Sahara during the Neolithic Subpluvial. Since that time, several cooling events have occurred, including:

- the Piora Oscillation

- the Middle Bronze Age Cold Epoch

- the Iron Age Cold Epoch

- cooling during the Dark Ages

- the Spörer Minimum

- the "Little Ice Age"

- the Dalton Minimum

- volcanic coolings such as from Laki in Iceland

- the phase of cooling c. 1940-1970, which led to global cooling hypotheses

In contrast, several warm periods have also taken place, and they include but are not limited to:

- the Older Peron during the late Holocene optimum

- a warm period during the apex of the Minoan civilization

- the Roman Warm Period

- the Medieval Warm Period

- the retreat of glaciers since 1850

- the "Modern Warming" during the 20th century

Certain effects have occurred during these cycles. For example, during the Medieval Warm Period, the American Midwest was in drought, including the Sand Hills of Nebraska which were active sand dunes. The black death plague of *Yersinia pestis* also occurred during Medieval temperature fluctuations, and may be related to changing climates.

Given that records of solar activity are accurate, solar activity may have contributed to part of the modern warming that peaked in the 1930s, in addition to the 60-year temperature cycles that result in roughly 0.5 °C of warming during the increasing temperature phase. However, solar cycles fail to account for warming observed since the 1980s to the present day. Events such as the opening of the Northwest Passage and recent record low ice minima of the modern Arctic shrinkage have not taken place for at least several centuries, as early explorers were all unable to make an Arctic crossing, even in summer. Shifts in biomes and habitat ranges are also unprecedented, occurring at rates that do not coincide with known climate oscillations. The extinction of many tropical amphibian species, especially in cloud forests, have been attributed to changing global temperatures, fungal disease and possible influence from unusually extreme phases of oceanic climate oscillations.

Climate State

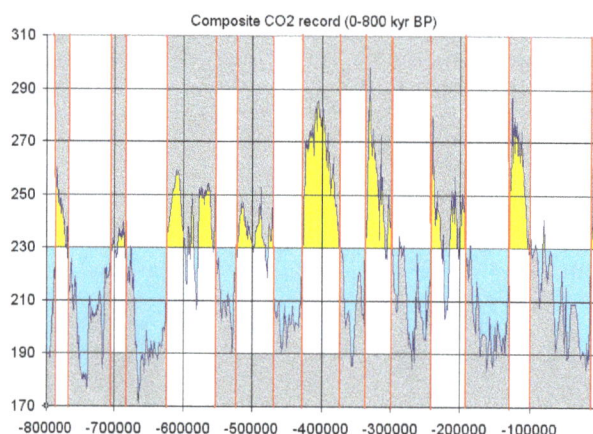

Ice core data of Glacial and Interglacial periods during the Quaternary.

Climate state describes a state of climate on Earth and similar terrestrial planets based on a thermal energy budget, such as the greenhouse or icehouse climate state.

The main climate state change is between periodical glacial and interglacial cycles in Earth history, studied from climate proxies. The climate system is responding to the current climate forcing and adjusts following climate sensitivity to reach a climate equilibrium, Earth's energy balance. Model simulations suggest that the current interglacial climate state will continue for at least another 100,000 years, due to CO_2 emissions - including complete deglaciation of the Northern Hemisphere.

General

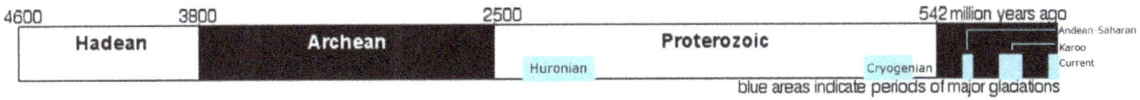

Timeline of glaciations (Ice Ages), shown in blue. The periods without glaciation are considered greenhouse states.

The orbital forcing from Milankovitch cycles is a periodical factor to determine Earth's energy budget and responsible for the glacial cycles on Earth, depending on the radiative equilibrium. Other factors include processes and change in geospheric systems. These include oceanic processes (such as oceanic circulation), biotic processes, variations in solar radiation received by Earth, plate tectonics, volcanism, albedo vegetation changes and human-induced alterations of the natural world.

The greenhouse has been the dominant state in Earth's past. Recovered ocean sediments of the past 120 million years contain evidence of the long-term transition from a greenhouse to icehouse climate state. A time when there are no glaciers on Earth is considered a greenhouse climate state. An ice age implies the presence of extensive ice sheets at Earth polar regions. The time during an Ice Age glacial period, when glaciers reach their maximum extent is referred to as icehouse climate state. There have been five known ice ages in the Earth's past, with the Earth experiencing currently an interglacial period (warming) during the present Quaternary Ice Age, identified as the "marine isotope stage 1" (MIS1) in the Holocene epoch (or recently the Anthropocene epoch).

The current climate state and evidence from the past of the climate system are important in determining the future evolution of climatic anomalies. Dansgaard–Oeschger events are considered switches between states of the climate system. Tipping points in the climate system describe thresholds, such as ice-albedo feedback which can cause abrupt climate change, and possibly leading to a new state. The climate state affects the formation processes of large volcanic provinces.

Climate States

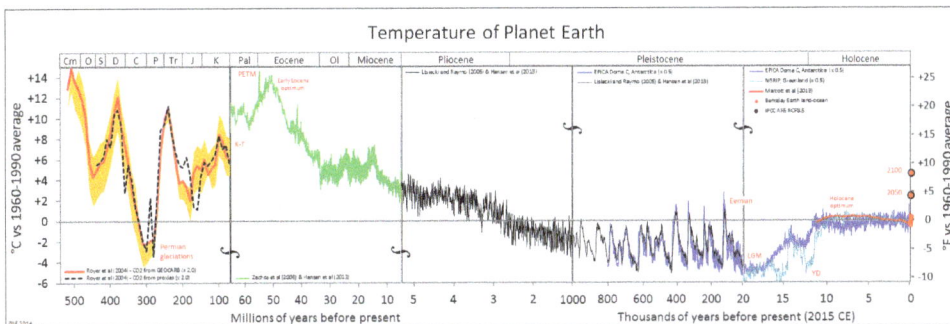

In the past the planet's climate has been fluctuating between two dominant states: the Greenhouse and the Icehouse state.

- The Huronian glaciation, is the first known glaciation in Earth's history, and lasted from 2400-2100 million years ago.

- The Cryogenian glaciation lasted from 850-635 million years ago.

- The Andean-Saharan glaciation lasted from 450–420 million years ago.

- The Karoo glaciation lasted from 360–260 million years ago.

- The Quaternary glaciation is the current glaciation period and begun 2.58 million years ago.

In the past, weathering of silicate rocks sequestered CO_2, a negative feedback loop which maintained Earth's climate within a habitable range over millions of years. When atmospheric CO_2 concentration rise, temperature and precipitation increase and thereby enhance chemical weathering. The last time global temperature reached a long-term maximum was during the Early Eocene Climatic Optimum, 51–53 million years ago. Only over the past 34 million years have CO_2 concentrations been low, temperatures relatively cool, and the poles glaciated. This long- term shift in Earth's climatic state resulted, in part from differences in volcanic emissions, which were particularly high during parts of the Palaeocene and Eocene epochs (about 40–60 million years ago) but have diminished since then.

Hothouse

The Hothouse climate state is postulated as a runaway climate change state, which might have happened on Venus. Ongoing research determines if such a state is possible on Earth.

Conceivable levels of human-made climate forcing could yield the *low-end* runaway greenhouse. A forcing of 12–16 $W m^{-2}$, which would require CO_2 to increase by a factor of 8–16 times, if the forcing were due only to CO_2 change, would raise the global mean temperature by 16–24 °C with much larger polar warming. That would melt all the ice on the planet, and probably thaw methane hydrates and scorch carbon from global peat deposits and tropical forests. This forcing would not produce the extreme Venus-like baked-crust greenhouse state, which cannot be reached until the ocean is lost to space. A warming of 16–24 °C produces a moderately moist greenhouse, with water vapour increasing to about 1% of the atmosphere's mass, thus increasing the rate of hydrogen escape to space. However, if the forcing is by fossil fuel CO_2, the weathering process would remove the excess atmospheric CO_2 on a time scale of 10^4–10^5 years, well before the ocean is significantly depleted. Baked-crust hothouse conditions on the Earth require a large long-term forcing that is unlikely to occur until the sun brightens by a few tens of per cent, which will take a few billion years.

Hansen et al. 2013 suggests that the Earth could become in large parts uninhabitable and noted that this may not even require burning all of fossil fuels, because of higher climate sensitivity (3–4 °C or 5.4–7.2 °F) based on a 550ppm scenario.

Icehouse

Snowball Earth describes the Icehouse climate state during the Neoproterozoic which caused glaciation from the planet's poles to the Equator.

State Changes

Based on climate proxies paleoclimatologists study the different climate states originating from glaciation. In climate science distinction is made between the *background state* (Today), the *initial state*, the *equilibrium state* and the paleo state, when considering climate sensitivity and climate forcing. In order to understand future climate projections, interactions of feedbacks and thus the calculated climate sensitivity with the background climate state has become a top priority in climate science. Feedbacks in the climate system, and thus climate sensitivity, may depend in an unknown non-linear manner on the climate state before perturbation ('background climate state') and on type of forcing. However, decadal to inter-decadal climate variability can be interpreted from past variability, and the identification of the involved dynamical processes can help to understand the chain of events which are characteristic for shifts in the climate state, today.

References

- Bralower, T.J.; Premoli Silva, I.; Malone, M.J. (2006). Leg 198 Synthesis : A Remarkable 120-m.y. Record of Climate and Oceanography from Shatsky Rise, Northwest Pacific Ocean. Proceedings of the Ocean drilling program. p. 47. doi:10.2973/odp.proc.ir.198.2002. ISSN 1096-2158. Retrieved April 9, 2014.

- Scafetta, Nicola (May 15, 2010). "Empirical evidence for a celestial origin of the climate oscillations" (PDF). Journal of Atmospheric and Solar-Terrestrial Physics. 72: 951–970. arXiv:1005.4639. Bibcode:2010JASTP..72..951S. doi:10.1016/j.jastp.2010.04.015. Retrieved 20 July 2011.

Global Warming: An Integrated Study

With the failure to reduce greenhouse gas emissions, scientists have turned to the study of present and possible future disasters as well as extreme weather that certain regions of the world has encountered. This chapter is an overview of the subject matter incorporating all the major aspects of global warming.

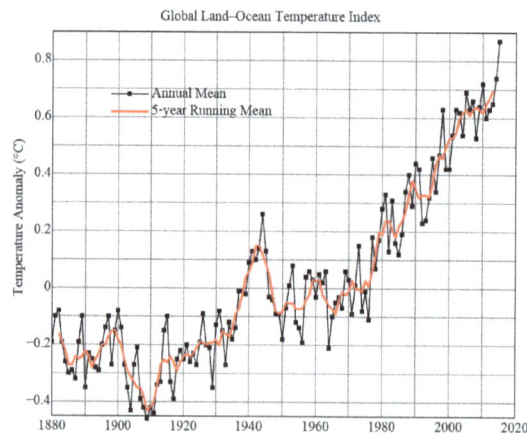

Global Warming

Global Land–Ocean Temperature Index

Global mean surface temperature change from 1880 to 2015, relative to the 1951–1980 mean. The black line is the annual mean and the red line is the 5-year running mean. Source: NASA GISS.

Global warming and climate change are terms for the observed century-scale rise in the average temperature of the Earth's climate system and its related effects. Multiple lines of scientific evidence show that the climate system is warming. Although the increase of near-surface atmospheric temperature is the measure of global warming often reported in the popular press, most of the additional energy stored in the climate system since 1970 has gone into the oceans. The rest has melted ice and warmed the continents and atmosphere.[a] Many of the observed changes since the 1950s are unprecedented over tens to thousands of years.

Scientific understanding of global warming is increasing. The Intergovernmental Panel on Climate Change (IPCC) reported in 2014 that scientists were more than 95% certain that global warming is mostly being caused by human (anthropogenic) activities, mainly increasing concentrations of greenhouse gases such as carbon dioxide (CO_2). Human-made carbon dioxide continues to increase above levels not seen in hundreds of thousands of years. Currently, about half of the carbon dioxide released from the burning of fossil fuels remains in the atmosphere. The rest is absorbed by vegetation and the oceans. Climate model projections summarized in the report indicated that during the 21st century the global surface temperature is likely to rise a further 0.3 to 1.7 °C (0.5 to 3.1 °F) for their lowest emissions scenario and 2.6 to 4.8 °C (4.7 to 8.6 °F) for the highest emissions scenario. These findings have been recognized by the national science academies of the major industrialized nations and are not disputed by any scientific body of national or international standing.

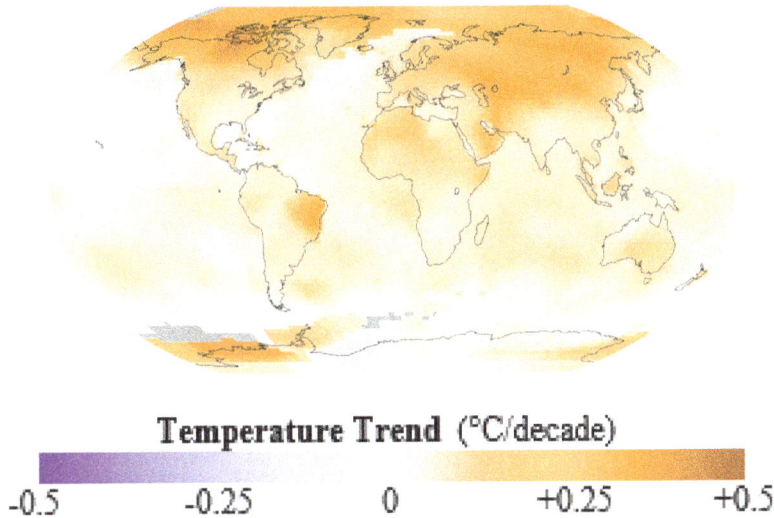

Temperature Trend (°C/decade)

-0.5   -0.25   0   +0.25   +0.5

World map showing surface temperature trends (°C per decade) between 1950 and 2014. Source: NASA GISS.

Future climate change and associated impacts will differ from region to region around the globe. Anticipated effects include warming global temperature, rising sea levels, changing precipitation, and expansion of deserts in the subtropics. Warming is expected to be greater over land than over the oceans and greatest in the Arctic, with the continuing retreat of glaciers, permafrost and sea ice. Other likely changes include more frequent extreme weather events including heat waves, droughts, heavy rainfall with floods and heavy snowfall; ocean acidification; and species extinctions due to shifting temperature regimes. Effects significant to humans include the threat to food security from decreasing crop yields and the abandonment of populated areas due to rising sea levels. Because the climate system has a large "inertia" and CO_2 will stay in the atmosphere for a long time, many of these effects will not only exist for decades or centuries, but will persist for tens of thousands of years.

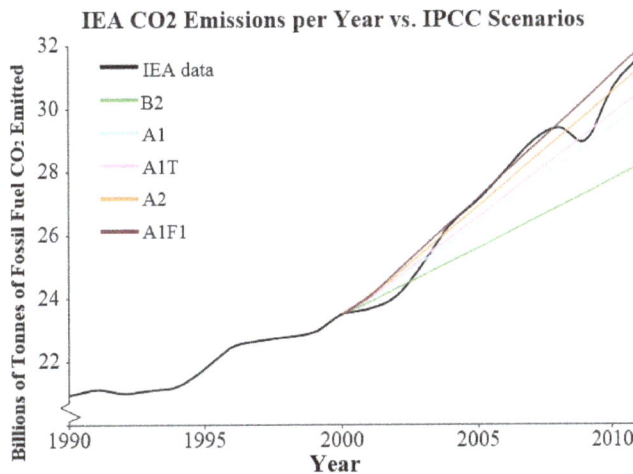

Fossil fuel related carbon dioxide (CO_2) emissions compared to five of the IPCC's "SRES" emissions scenarios, published in 2000. The dips are related to global recessions. Image source: Skeptical Science.

Possible societal responses to global warming include mitigation by emissions reduction, adaptation to its effects, building systems resilient to its effects, and possible future climate engineering. Most countries are parties to the United Nations Framework Convention on Climate Change (UN-FCCC), whose ultimate objective is to prevent dangerous anthropogenic climate change. Parties to the UNFCCC have agreed that deep cuts in emissions are required and that global warming should be limited to well below 2.0 °C (3.6 °F) relative to pre-industrial levels,[c] with efforts made to limit warming to 1.5 °C (2.7 °F).

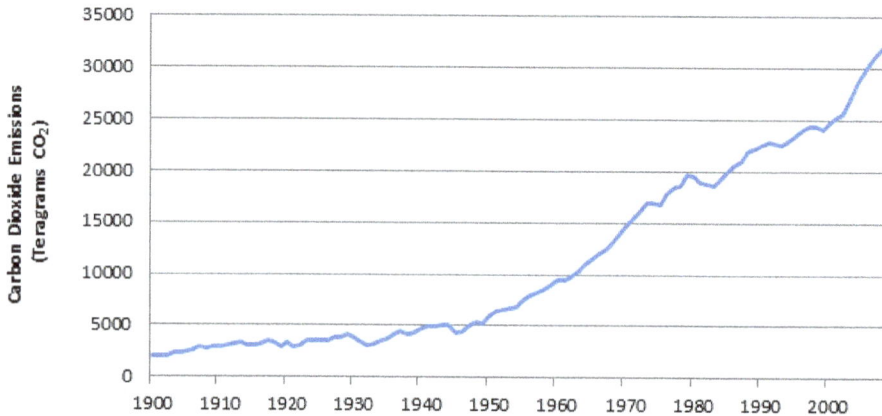

Fossil fuel related carbon dioxide emissions over the 20th century. Image source: EPA.

Public reactions to global warming and concern about its effects are also increasing. A global 2015 Pew Research Center report showed a median of 54% consider it "a very serious problem". There are significant regional differences, with Americans and Chinese (whose economies are responsible for the greatest annual CO2 emissions) among the least concerned.

Observed Temperature Changes

Energy change inventory, 1971-2010

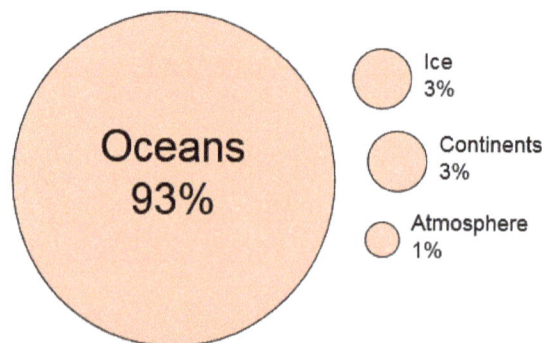

Earth has been in *radiative imbalance* since at least the 1970s, where less energy leaves the atmosphere than enters it. Most of this extra energy has been absorbed by the oceans. It is very likely that human activities substantially contributed to this increase in ocean heat content.

The global average (land and ocean) surface temperature shows a warming of 0.85 [0.65 to 1.06] °C in the period 1880 to 2012, based on multiple independently produced datasets. Earth's average surface temperature rose by 0.74±0.18 °C over the period 1906–2005.

The rate of warming almost doubled for the last half of that period (0.13±0.03 °C per decade, versus 0.07±0.02 °C per decade).

Reconstructed Temperature

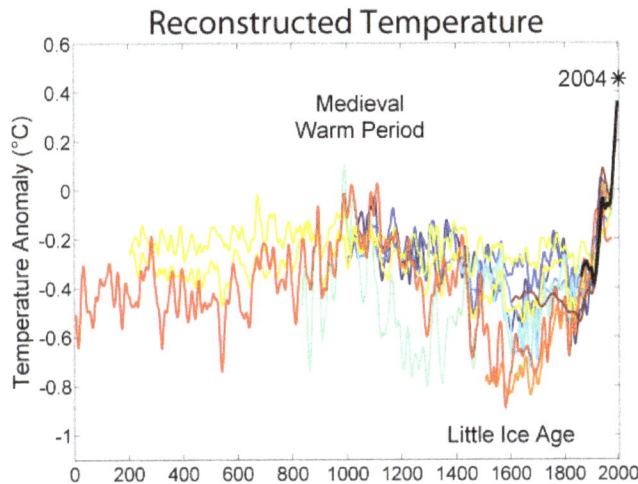

Two millennia of mean surface temperatures according to different reconstructions from climate proxies, each smoothed on a decadal scale, with the instrumental temperature record overlaid in black.

The average temperature of the lower troposphere has increased between 0.13 and 0.22 °C (0.23 and 0.40 °F) per decade since 1979, according to satellite temperature measurements. Climate proxies show the temperature to have been relatively stable over the one or two thousand years before 1850, with regionally varying fluctuations such as the Medieval Warm Period and the Little Ice Age.

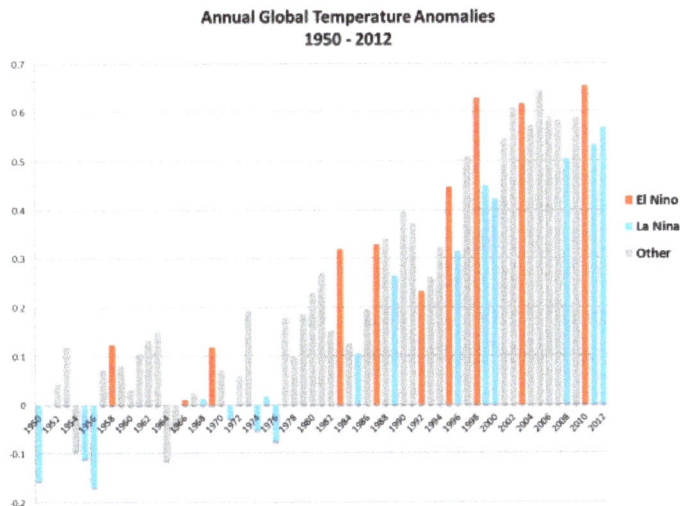

NOAA graph of Global Annual Temperature Anomalies 1950–2012, showing the El Niño Southern Oscillation

The warming that is evident in the instrumental temperature record is consistent with a wide range of observations, as documented by many independent scientific groups. Examples include sea level rise, widespread melting of snow and land ice, increased heat content of the oceans, increased humidity, and the earlier timing of spring events, e.g., the flowering of plants. The probability that these changes could have occurred by chance is virtually zero.

Trends

Temperature changes vary over the globe. Since 1979, land temperatures have increased about twice as fast as ocean temperatures (0.25 °C per decade against 0.13 °C per decade). Ocean temperatures increase more slowly than land temperatures because of the larger effective heat capacity of the oceans and because the ocean loses more heat by evaporation. Since the beginning of industrialisation the temperature difference between the hemispheres has increased due to melting of sea ice and snow in the North. Average arctic temperatures have been increasing at almost twice the rate of the rest of the world in the past 100 years; however arctic temperatures are also highly variable. Although more greenhouse gases are emitted in the Northern than Southern Hemisphere this does not contribute to the difference in warming because the major greenhouse gases persist long enough to mix between hemispheres.

The thermal inertia of the oceans and slow responses of other indirect effects mean that climate can take centuries or longer to adjust to changes in forcing. One climate commitment study concluded that if greenhouse gases were stabilized at year 2000 levels, surface temperatures would still increase by about one-half degree Celsius, and another found that if they were stabilized at 2005 levels surface warming could exceed a whole degree Celsius. Some of this surface warming will be driven by past natural forcings which are still seeking equilibrium in the climate system. One study using a highly simplified climate model indicates these past natural forcings may account for as much as 64% of the committed 2050 surface warming and their influence will fade with time compared to the human contribution.

Global temperature is subject to short-term fluctuations that overlay long-term trends and can temporarily mask them. The relative stability in surface temperature from 2002 to 2009, which has been dubbed the global warming hiatus by the media and some scientists, is consistent with such an episode. 2015 updates to account for differing methods of measuring ocean surface temperature measurements show a positive trend over the recent decade.

Warmest Years

15 of the top 16 warmest years have occurred since 2000. While record-breaking years can attract considerable public interest, individual years are less significant than the overall trend. So some climatologists have criticized the attention that the popular press gives to "warmest year" statistics; for example, Gavin Schmidt stated "the long-term trends or the expected sequence of records are far more important than whether any single year is a record or not."

2015 was not only the warmest year on record, it broke the record by the largest margin by which the record has been broken. 2015 was the 39th consecutive year with above-average temperatures. Ocean oscillations like El Niño Southern Oscillation (ENSO) can affect global average temperatures, for example, 1998 temperatures were significantly enhanced by strong El Niño conditions. 1998 remained the warmest year until 2005 and 2010 and the temperature of both of these years was enhanced by El Niño periods. The large margin by which 2015 is the warmest year is also attributed to another strong El Niño. However, 2014 was ENSO neutral. According to NOAA and NASA, 2015 had the warmest respective months on record for 10 out of the 12 months. The average temperature around the globe was 1.62°F (0.90°C) or 20% above the twentieth century average. In a first, December 2015 was also the first month to ever reach a temperature 2 degrees Fahrenheit above normal for the planet. As of July 2016, for the 15th consecutive month, the global land and ocean temperature departure from average was the highest since global temperature records began in 1880. This marks the longest such streak in NOAA's 137 years of record keeping. In addition, NOAA reported that July of 2016 was the hottest month on record.

Initial Causes Of Temperature Changes (External Forcings)

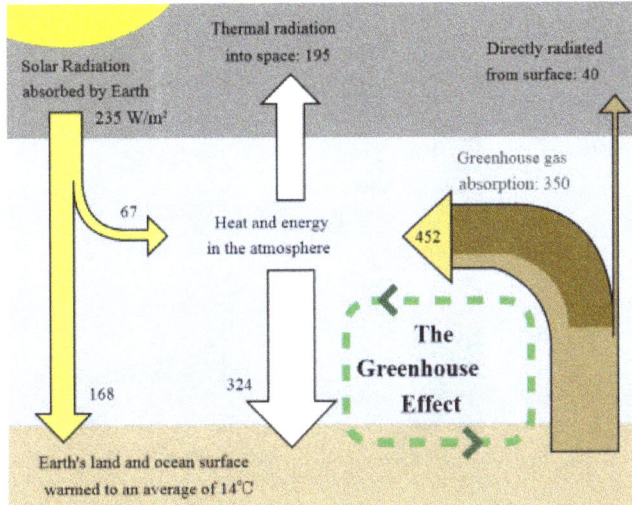

Greenhouse effect schematic showing energy flows between space, the atmosphere, and Earth's surface. Energy exchanges are expressed in watts per square metre (W/m²).

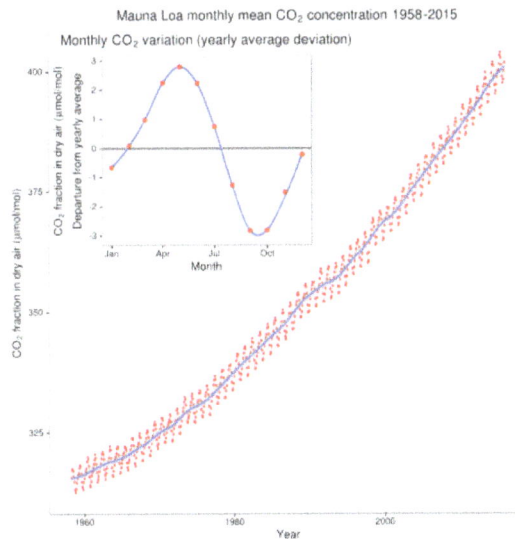

This graph, known as the Keeling Curve, documents the increase of atmospheric carbon dioxide concentrations from 1958–2015. Monthly CO_2 measurements display seasonal oscillations in an upward trend; each year's maximum occurs during the Northern Hemisphere's late spring, and declines during its growing season as plants remove some atmospheric CO_2.

The climate system can warm or cool in response to changes in *external forcings*. These are "external" to the climate system but not necessarily external to Earth. Examples of external forcings include changes in atmospheric composition (e.g., increased concentrations of greenhouse gases), solar luminosity, volcanic eruptions, and variations in Earth's orbit around the Sun.

Greenhouse Gases

The greenhouse effect is the process by which absorption and emission of infrared radiation by gases in a planet's atmosphere warm its lower atmosphere and surface. It was proposed by Joseph Fourier in 1824, discovered in 1860 by John Tyndall, was first investigated quantitatively by Svante Arrhenius in 1896, and was developed in the 1930s through 1960s by Guy Stewart Callendar.

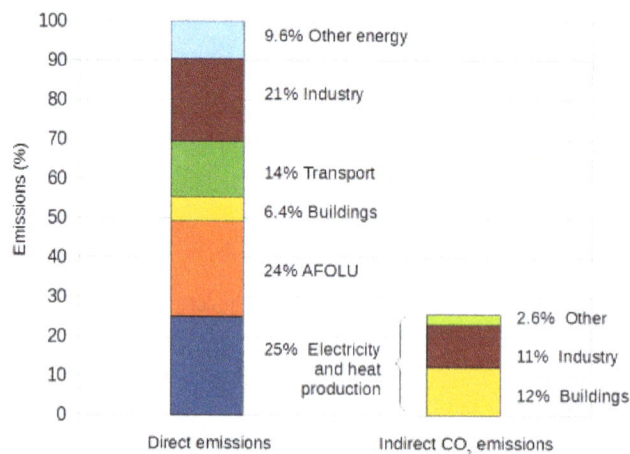

Annual world greenhouse gas emissions, in 2010, by sector.

Percentage share of global cumulative energy-related CO_2 emissions between 1751 and 2012 across different regions.

On Earth, naturally occurring amounts of greenhouse gases cause air temperature near the surface to be about 33 °C (59 °F) warmer than it would be in their absence.[d] Without the Earth's atmosphere, the Earth's average temperature would be well below the freezing temperature of water. The major greenhouse gases are water vapour, which causes about 36–70% of the greenhouse effect; carbon dioxide (CO_2), which causes 9–26%; methane (CH_4), which causes 4–9%; and ozone (O_3), which causes 3–7%. Clouds also affect the radiation balance through cloud forcings similar to greenhouse gases.

Human activity since the Industrial Revolution has increased the amount of greenhouse gases in the atmosphere, leading to increased radiative forcing from CO_2, methane, tropospheric ozone, CFCs and nitrous oxide. According to work published in 2007, the concentrations of CO_2 and methane have increased by 36% and 148% respectively since 1750. These levels are much higher than at any time during the last 800,000 years, the period for which reliable data has been extracted from ice cores. Less direct geological evidence indicates that CO_2 values higher than this were last seen about 20 million years ago.

Fossil fuel burning has produced about three-quarters of the increase in CO_2 from human activity over the past 20 years. The rest of this increase is caused mostly by changes in land-use, particularly deforestation. Another significant non-fuel source of anthropogenic CO_2 emissions is the calcination of limestone for clinker production, a chemical process which releases CO_2. Estimates of global CO_2 emissions in 2011 from fossil fuel combustion, including cement production and gas flaring, was 34.8 billion tonnes (9.5 ± 0.5 PgC), an increase of 54% above emissions in 1990. Coal burning was responsible for 43% of the total emissions, oil 34%, gas 18%, cement 4.9% and gas flaring 0.7%

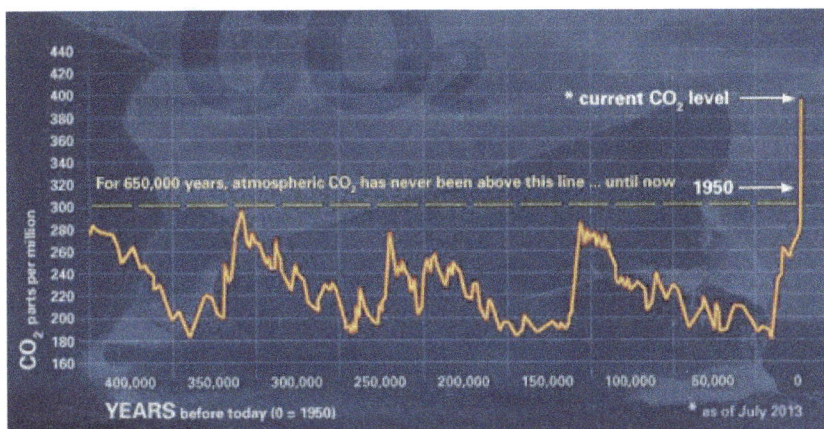

Atmospheric CO_2 concentration from 650,000 years ago to near present, using ice core proxy data and direct measurements.

In May 2013, it was reported that readings for CO_2 taken at the world's primary benchmark site in Mauna Loa surpassed 400 ppm. According to professor Brian Hoskins, this is likely the first time CO_2 levels have been this high for about 4.5 million years. Monthly global CO_2 concentrations exceeded 400 ppm in March 2015, probably for the first time in several million years. On 12 November 2015, NASA scientists reported that human-made carbon dioxide continues to increase above levels not seen in hundreds of thousands of years: currently, about half of the carbon dioxide released from the burning of fossil fuels is not absorbed by vegetation and the oceans and remains in the atmosphere.

Over the last three decades of the twentieth century, gross domestic product per capita and population growth were the main drivers of increases in greenhouse gas emissions. CO_2 emissions are continuing to rise due to the burning of fossil fuels and land-use change. Emissions can be attributed to different regions. Attributions of emissions due to land-use change are subject to considerable uncertainty.

Emissions scenarios, estimates of changes in future emission levels of greenhouse gases, have been projected that depend upon uncertain economic, sociological, technological, and natural developments. In most scenarios, emissions continue to rise over the century, while in a few, emissions are reduced. Fossil fuel reserves are abundant, and will not limit carbon emissions in the 21st century. Emission scenarios, combined with modelling of the carbon cycle, have been used to produce estimates of how atmospheric concentrations of greenhouse gases might change in the future. Using the six IPCC SRES "marker" scenarios, models suggest that by the year 2100, the atmospheric concentration of CO_2 could range between 541 and 970 ppm. This is 90–250% above the concentration in the year 1750.

The popular media and the public often confuse global warming with ozone depletion, i.e., the destruction of stratospheric ozone (e.g., the ozone layer) by chlorofluorocarbons. Although there are a few areas of linkage, the relationship between the two is not strong. Reduced stratospheric ozone has had a slight cooling influence on surface temperatures, while increased tropospheric ozone has had a somewhat larger warming effect.

Aerosols and Soot

Ship tracks can be seen as lines in these clouds over the Atlantic Ocean on the east coast of the United States. Atmospheric particles from these and other sources could have a large effect on climate through the aerosol indirect effect.

Global dimming, a gradual reduction in the amount of global direct irradiance at the Earth's surface, was observed from 1961 until at least 1990. Solid and liquid particles known as *aerosols*, produced by volcanoes and human-made pollutants, are thought to be the main cause of this dimming. They exert a cooling effect by increasing the reflection of incoming sunlight. The effects of the products of fossil fuel combustion – CO_2 and aerosols – have partially offset one another in recent decades, so that net warming has been due to the increase in non-CO_2 greenhouse gases such as methane. Radiative forcing due to aerosols is temporally limited due to the processes that remove aerosols from the atmosphere. Removal by clouds and precipitation gives tropospheric aerosols an atmospheric lifetime of only about a week, while stratospheric aerosols can remain for a few years. Carbon dioxide has a lifetime of a century or more, and as such, changes in aerosols will only delay climate changes due to carbon dioxide. Black carbon is second only to carbon dioxide for its contribution to global warming.

In addition to their direct effect by scattering and absorbing solar radiation, aerosols have indirect effects on the Earth's radiation budget. Sulfate aerosols act as cloud condensation nuclei and thus lead to clouds that have more and smaller cloud droplets. These clouds reflect solar radiation more efficiently than clouds with fewer and larger droplets, a phenomenon known as the Twomey effect. This effect also causes droplets to be of more uniform size, which reduces growth of raindrops and makes the cloud more reflective to incoming sunlight, known as the Albrecht effect. Indirect effects are most noticeable in marine stratiform clouds, and have very little radiative effect on convective clouds. Indirect effects of aerosols represent the largest uncertainty in radiative forcing.

Soot may either cool or warm Earth's climate system, depending on whether it is airborne or deposited. Atmospheric soot directly absorbs solar radiation, which heats the atmosphere and cools the surface. In isolated areas with high soot production, such as rural India, as much as 50% of surface warming due to greenhouse gases may be masked by atmospheric brown clouds. When deposited, especially on glaciers or on ice in arctic regions, the lower surface albedo can also directly heat the surface. The influences of atmospheric particles, including black carbon, are most pronounced in the tropics and sub-tropics, particularly in Asia, while the effects of greenhouse gases are dominant in the extratropics and southern hemisphere.

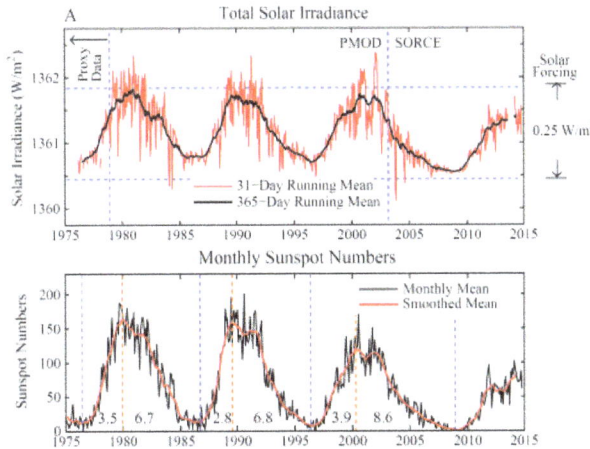

Changes in Total Solar Irradiance (TSI) and monthly sunspot numbers since the mid-1970s.

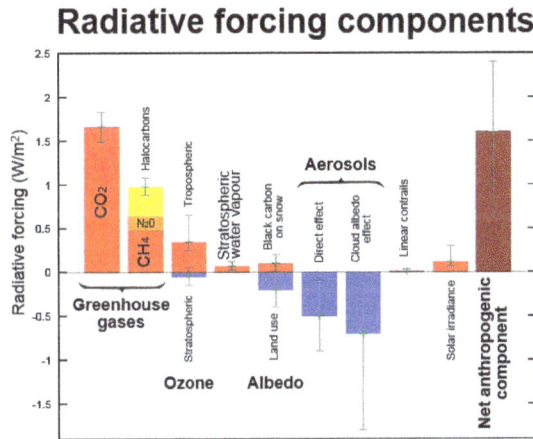

Contribution of natural factors and human activities to radiative forcing of climate change. Radiative forcing values are for the year 2005, relative to the pre-industrial era (1750). The contribution of solar irradiance to radiative forcing is 5% the value of the combined radiative forcing due to increases in the atmospheric concentrations of carbon dioxide, methane and nitrous oxide.

Solar Activity

Since 1978, solar irradiance has been measured by satellites. These measurements indicate that the Sun's radiative output has not increased since 1978, so the warming during the past 30 years cannot be attributed to an increase in solar energy reaching the Earth.

Climate models have been used to examine the role of the Sun in recent climate change. Models are unable to reproduce the rapid warming observed in recent decades when they only take into account variations in solar output and volcanic activity. Models are, however, able to simulate the observed 20th century changes in temperature when they include all of the most important external forcings, including human influences and natural forcings.

Another line of evidence against solar variations having caused recent climate change comes from looking at how temperatures at different levels in the Earth's atmosphere have changed. Models and observations show that greenhouse warming results in warming of the lower atmosphere (the troposphere) but cooling of the upper atmosphere (the stratosphere). Depletion of the ozone layer by chemical refrigerants has also resulted in a strong cooling effect in the stratosphere. If solar variations were responsible for observed warming, warming of both the troposphere and stratosphere would be expected.

Variations in Earth's Orbit

The tilt of the Earth's axis and the shape of its orbit around the Sun vary slowly over tens of thousands of years and are a natural source of climate change, by changing the seasonal and latitudinal distribution of solar insolation.

During the last few thousand years, this phenomenon contributed to a slow cooling trend at high latitudes of the Northern Hemisphere during summer, a trend that was reversed by greenhouse-gas-induced warming during the 20th century.

Variations in orbital cycles may initiate a new glacial period in the future, though the timing of this depends on greenhouse gas concentrations as well as the orbital forcing. A new glacial period is not expected within the next 50,000 years if atmospheric CO_2 concentration remains above 300 ppm.

Feedback

Sea ice, shown here in Nunavut, in northern Canada, reflects more sunshine, while open ocean absorbs more, accelerating melting.

The climate system includes a range of *feedbacks*, which alter the response of the system to changes in external forcings. Positive feedbacks increase the response of the climate system to an initial forcing, while negative feedbacks reduce it.

There are a range of feedbacks in the climate system, including water vapour, changes in ice-albedo (snow and ice cover affect how much the Earth's surface absorbs or reflects incoming sunlight), clouds, and changes in the Earth's carbon cycle (e.g., the release of carbon from soil). The main negative feedback is the energy the Earth's surface radiates into space as infrared radiation. According to the Stefan-Boltzmann law, if the absolute temperature (as measured in kelvin) doubles,[e] radiated energy increases by a factor of 16 (2 to the 4th power).

Feedbacks are an important factor in determining the sensitivity of the climate system to increased atmospheric greenhouse gas concentrations. Other factors being equal, a higher *climate sensitivity* means that more warming will occur for a given increase in greenhouse gas forcing. Uncertainty over the effect of feedbacks is a major reason why different climate models project different magnitudes of warming for a given forcing scenario. More research is needed to understand the role of clouds and carbon cycle feedbacks in climate projections.

The IPCC projections previously mentioned span the "likely" range (greater than 66% probability, based on expert judgement) for the selected emissions scenarios. However, the IPCC's projections do not reflect the full range of uncertainty. The lower end of the "likely" range appears to be better constrained than the upper end.

Climate Models

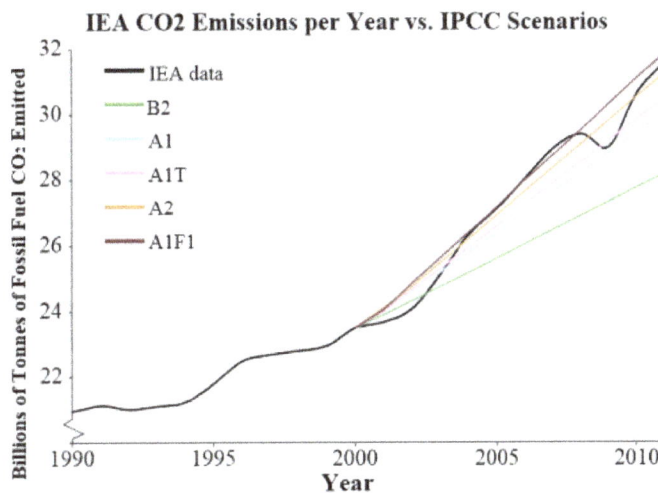

IEA CO2 Emissions per Year vs. IPCC Scenarios

Calculations of global warming prepared in or before 2001 from a range of climate models under the SRES A2 emissions scenario, which assumes no action is taken to reduce emissions and regionally divided economic development.

Projected change in annual mean surface air temperature from the late 20th century to the middle 21st century, based on a medium emissions scenario (SRES A1B). This scenario assumes that no future policies are adopted to limit greenhouse gas emissions. Image credit: NOAA GFDL.

A climate model is a representation of the physical, chemical and biological processes that affect the climate system. Such models are based on scientific disciplines such as fluid dynamics and thermodynamics as well as physical processes such as radiative transfer. The models may be used to predict a range of variables such as local air movement, temperature, clouds, and other atmospheric properties; ocean temperature, salt content, and circulation; ice cover on land and sea; the transfer of heat and moisture from soil and vegetation to the atmosphere; and chemical and biological processes, among others.

Although researchers attempt to include as many processes as possible, simplifications of the actual climate system are inevitable because of the constraints of available computer power and limitations in knowledge of the climate system. Results from models can also vary due to different greenhouse gas inputs and the model's climate sensitivity. For example, the uncertainty in IPCC's 2007 projections is caused by (1) the use of multiple models with differing sensitivity to greenhouse gas concentrations, (2) the use of differing estimates of humanity's future greenhouse gas emissions, (3) any additional emissions from climate feedbacks that were not included in the models IPCC used to prepare its report, i.e., greenhouse gas releases from permafrost.

The models do not assume the climate will warm due to increasing levels of greenhouse gases. Instead the models predict how greenhouse gases will interact with radiative transfer and other physical processes. Warming or cooling is thus a result, not an assumption, of the models.

Clouds and their effects are especially difficult to predict. Improving the models' representation of clouds is therefore an important topic in current research. Another prominent research topic is expanding and improving representations of the carbon cycle.

Models are also used to help investigate the causes of recent climate change by comparing the observed changes to those that the models project from various natural and human causes. Although these models do not unambiguously attribute the warming that occurred from approximately 1910 to 1945 to either natural variation or human effects, they do indicate that the warming since 1970 is dominated by anthropogenic greenhouse gas emissions.

The physical realism of models is tested by examining their ability to simulate contemporary or past climates. Climate models produce a good match to observations of global temperature changes over the last century, but do not simulate all aspects of climate. Not all effects of global warming are accurately predicted by the climate models used by the IPCC. Observed Arctic shrinkage has been faster than that predicted. Precipitation increased proportionally to atmospheric humidity, and hence significantly faster than global climate models predict. Since 1990, sea level has also risen considerably faster than models predicted it would.

Observed and Expected Environmental Effects

Anthropogenic forcing has likely contributed to some of the observed changes, including sea level rise, changes in climate extremes (such as the number of warm and cold days), declines in Arctic sea ice extent, glacier retreat, and greening of the Sahara.

During the 21st century, glaciers and snow cover are projected to continue their widespread retreat. Projections of declines in Arctic sea ice vary. Recent projections suggest that Arctic summers could be ice-free (defined as ice extent less than 1 million square km) as early as 2025-2030.

"Detection" is the process of demonstrating that climate has changed in some defined statistical sense, without providing a reason for that change. Detection does not imply attribution of the detected change to a particular cause. "Attribution" of causes of climate change is the process of establishing the most likely causes for the detected change with some defined level of confidence. Detection and attribution may also be applied to observed changes in physical, ecological and social systems.

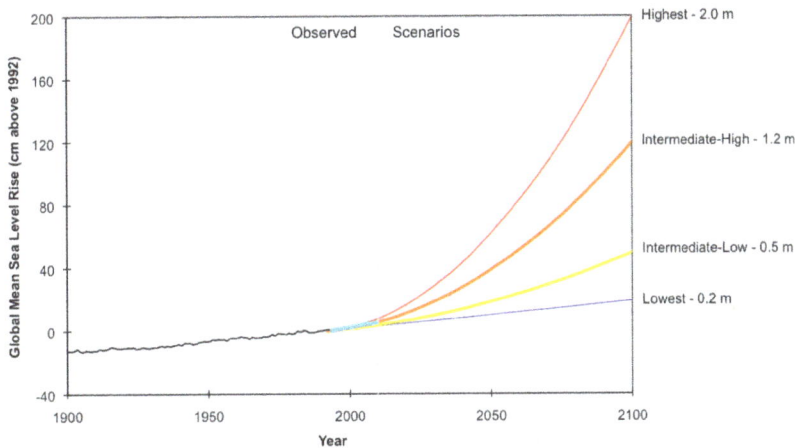

Projections of global mean sea level rise by Parris and others. Probabilities have not been assigned to these projections. Therefore, none of these projections should be interpreted as a "best estimate" of future sea level rise. Image credit: NOAA.

Extreme Weather

Changes in regional climate are expected to include greater warming over land, with most warming at high northern latitudes, and least warming over the Southern Ocean and parts of the North Atlantic Ocean.

Future changes in precipitation are expected to follow existing trends, with reduced precipitation over subtropical land areas, and increased precipitation at subpolar latitudes and some equatorial regions. Projections suggest a probable increase in the frequency and severity of some extreme weather events, such as heat waves.

A 2015 study published in *Nature Climate Change*, states:

> About 18% of the moderate daily precipitation extremes over land are attributable to the observed temperature increase since pre-industrial times, which in turn primarily results from human influence. For 2 °C of warming the fraction of precipitation extremes attributable to human influence rises to about 40%. Likewise, today about 75% of the moderate daily hot extremes over land are attributable to warming. It is the most rare and extreme events for which the largest fraction is anthropogenic, and that contribution increases nonlinearly with further warming.

Data analysis of extreme events from 1960 till 2010 suggests that droughts and heat waves appear simultaneously with increased frequency. Extremely wet or dry events within the monsoon period have increased since 1980.

Sea level rise

The sea level rise since 1993 has been estimated to have been on average 2.6 mm and 2.9 mm per year ± 0.4 mm. Additionally, sea level rise has accelerated from 1995 to 2015. Over the 21st century, the IPCC projects for a high emissions scenario, that global mean sea level could rise by 52–98 cm. The IPCC's projections are conservative, and may underestimate future sea level rise. Other estimates suggest that for the same period, global mean sea level could rise by 0.2 to 2.0 m (0.7–6.6 ft), relative to mean sea level in 1992.

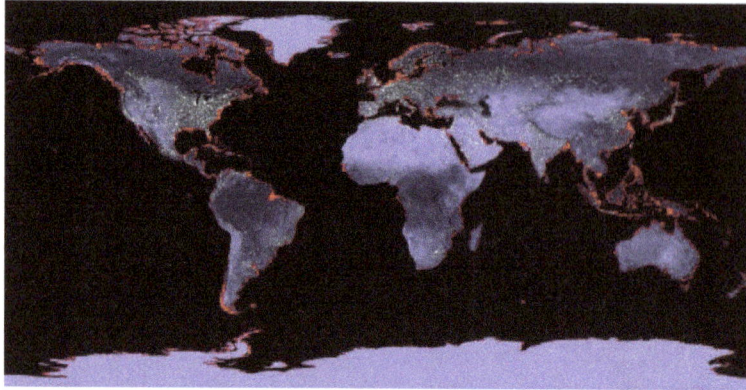

Map of the Earth with a six-metre sea level rise represented in red.

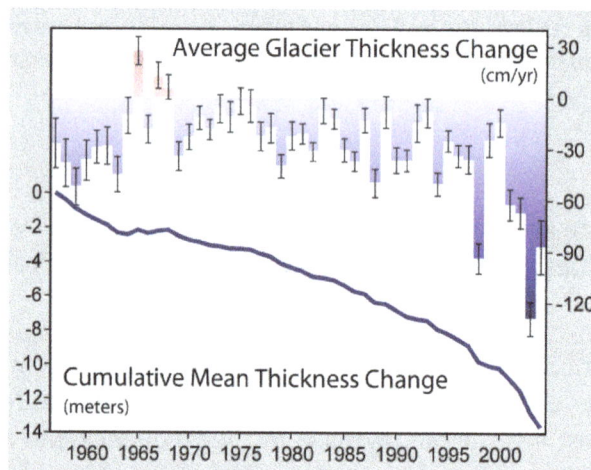

Sparse records indicate that glaciers have been retreating since the early 1800s. In the 1950s measurements began that allow the monitoring of glacial mass balance, reported to the World Glacier Monitoring Service (WGMS) and the National Snow and Ice Data Center (NSIDC).

Widespread coastal flooding would be expected if several degrees of warming is sustained for millennia. For example, sustained global warming of more than 2 °C (relative to pre-industrial levels) could lead to eventual sea level rise of around 1 to 4 m due to thermal expansion of sea water and the melting of glaciers and small ice caps. Melting of the Greenland ice sheet could contribute an additional 4 to 7.5 m over many thousands of years. It has been estimated that we are already committed to a sea-level rise of approximately 2.3 metres for each degree of temperature rise within the next 2,000 years.

Warming beyond the 2 °C target would potentially lead to rates of sea-level rise dominated by ice loss from Antarctica. Continued CO_2 emissions from fossil sources could cause additional tens of metres of sea level rise, over the next millennia and eventually ultimately eliminate the entire Antarctic ice sheet, causing about 58 metres of sea level rise.

Ecological Systems

In terrestrial ecosystems, the earlier timing of spring events, as well as poleward and upward shifts in plant and animal ranges, have been linked with high confidence to recent warming. Future climate change is expected to affect particular ecosystems, including tundra, mangroves, and coral reefs. It is expected that most ecosystems will be affected by higher atmospheric CO_2 levels, combined with higher global temperatures. Overall, it is expected that climate change will result in the extinction of many species and reduced diversity of ecosystems.

Increases in atmospheric CO_2 concentrations have led to an increase in ocean acidity. Dissolved CO_2 increases ocean acidity, measured by lower pH values. Between 1750 and 2000, surface-ocean pH has decreased by ≈ 0.1, from ≈ 8.2 to ≈ 8.1. Surface-ocean pH has probably not been below ≈ 8.1 during the past 2 million years. Projections suggest that surface-ocean pH could decrease by an additional 0.3–0.4 units by 2100. Future ocean acidification could threaten coral reefs, fisheries, protected species, and other natural resources of value to society.

Ocean deoxygenation is projected to increase hypoxia by 10%, and triple suboxic waters (oxygen concentrations 98% less than the mean surface concentrations), for each 1 °C of upper ocean warming.

Long-term Effects

On the timescale of centuries to millennia, the magnitude of global warming will be determined primarily by anthropogenic CO_2 emissions. This is due to carbon dioxide's very long lifetime in the atmosphere.

Stabilizing the global average temperature would require large reductions in CO_2 emissions, as well as reductions in emissions of other greenhouse gases such as methane and nitrous oxide. Emissions of CO_2 would need to be reduced by more than 80% relative to their peak level. Even if this were achieved, global average temperatures would remain close to their highest level for many centuries.

Long-term effects also include a response from the Earth's crust, due to ice melting and deglaciation, in a process called post-glacial rebound, when land masses are no longer depressed by the weight of ice. This could lead to landslides and increased seismic and volcanic activities. Tsunamis could be generated by submarine landslides caused by warmer ocean water thawing ocean-floor permafrost or releasing gas hydrates. Some world regions, such as the French Alps, already show signs of an increase in landslide frequency.

Large-scale and Abrupt Impacts

Climate change could result in global, large-scale changes in natural and social systems. Examples include the possibility for the Atlantic Meridional Overturning Circulation to slow- or shutdown, which in the instance of a shutdown would change weather in Europe and North America considerably, ocean acidification caused by increased atmospheric concentrations of carbon dioxide, and the long-term melting of ice sheets, which contributes to sea level rise.

Some large-scale changes could occur abruptly, i.e., over a short time period, and might also be irreversible. Examples of abrupt climate change are the rapid release of methane and carbon dioxide from permafrost, which would lead to amplified global warming, or the shutdown of thermohaline circulation. Scientific understanding of abrupt climate change is generally poor. The probability of abrupt change for some climate related feedbacks may be low. Factors that may increase the probability of abrupt climate change include higher magnitudes of global warming, warming that occurs more rapidly, and warming that is sustained over longer time periods.

Observed and Expected Effects on Social Systems

The effects of climate change on human systems, mostly due to warming or shifts in precipitation patterns, or both, have been detected worldwide. Production of wheat and maize globally has been impacted by climate change. While crop production has increased in some mid-latitude regions such as the UK and Northeast China, economic losses due to extreme weather events have increased globally. There has been a shift from cold- to heat-related mortality in some regions as a result of warming. Livelihoods of indigenous peoples of the Arctic have been altered by climate change, and there is emerging evidence of climate change impacts on livelihoods of indigenous peoples in other regions. Regional impacts of climate change are now observable at more locations than before, on all continents and across ocean regions.

The future social impacts of climate change will be uneven. Many risks are expected to increase with higher magnitudes of global warming. All regions are at risk of experiencing negative impacts. Low-latitude, less developed areas face the greatest risk. A study from 2015 concluded that economic growth (gross domestic product) of poorer countries is much more impaired with projected future climate warming, than previously thought.

A meta-analysis of 56 studies concluded in 2014 that each degree of temperature rise will increase violence by up to 20%, which includes fist fights, violent crimes, civil unrest or wars.

Examples of impacts include:

- *Food*: Crop production will probably be negatively affected in low latitude countries, while effects at northern latitudes may be positive or negative. Global warming of around 4.6 °C relative to pre-industrial levels could pose a large risk to global and regional food security.

- *Health*: Generally impacts will be more negative than positive. Impacts include: the effects of extreme weather, leading to injury and loss of life; and indirect effects, such as undernutrition brought on by crop failures.

Habitat Inundation

In small islands and mega deltas, inundation as a result of sea level rise is expected to threaten vital infrastructure and human settlements. This could lead to issues of homelessness in countries with low-lying areas such as Bangladesh, as well as statelessness for populations in countries such as the Maldives and Tuvalu.

Economy

Estimates based on the IPCC A1B emission scenario from additional CO_2 and CH_4 greenhouse gases released from permafrost, estimate associated impact damages by US$43 trillion.

Infrastructure

Continued permafrost degradation will likely result in unstable infrastructure in Arctic regions, or Alaska before 2100. Thus, impacting roads, pipelines and buildings, as well as water distribution, and cause slope failures.

Possible Responses to Global Warming

Mitigation

The graph on the right shows three "pathways" to meet the UNFCCC's 2 °C target, labelled "global technology", "decentralized solutions", and "consumption change". Each pathway shows how various measures (e.g., improved energy efficiency, increased use of renewable energy) could contribute to emissions reductions. Image credit: PBL Netherlands Environmental Assessment Agency.

Mitigation of climate change are actions to reduce greenhouse gas emissions, or enhance the capacity of carbon sinks to absorb GHGs from the atmosphere. There is a large potential for future reductions in emissions by a combination of activities, including: energy conservation and increased energy efficiency; the use of low-carbon energy technologies, such as renewable energy, nuclear energy, and carbon capture and storage; and enhancing carbon sinks through, for example, reforestation and preventing deforestation. A 2015 report by Citibank concluded that transitioning to a low carbon economy would yield positive return on investments.

Near- and long-term trends in the global energy system are inconsistent with limiting global warming at below 1.5 or 2 °C, relative to pre-industrial levels. Pledges made as part of the Cancún agreements are broadly consistent with having a likely chance (66 to 100% probability) of limiting global warming (in the 21st century) at below 3 °C, relative to pre-industrial levels.

In limiting warming at below 2 °C, more stringent emission reductions in the near-term would allow for less rapid reductions after 2030. Many integrated models are unable to meet the 2 °C target if pessimistic assumptions are made about the availability of mitigation technologies.

Adaptation

Other policy responses include adaptation to climate change. Adaptation to climate change may be planned, either in reaction to or anticipation of climate change, or spontaneous, i.e., without government intervention. Planned adaptation is already occurring on a limited basis. The barriers, limits, and costs of future adaptation are not fully understood.

A concept related to adaptation is *adaptive capacity*, which is the ability of a system (human, natural or managed) to adjust to climate change (including climate variability and extremes) to moderate potential damages, to take advantage of opportunities, or to cope with consequences. Unmitigated climate change (i.e., future climate change without efforts to limit greenhouse gas emissions) would, in the long term, be likely to exceed the capacity of natural, managed and human systems to adapt.

Environmental organizations and public figures have emphasized changes in the climate and the risks they entail, while promoting adaptation to changes in infrastructural needs and emissions reductions.

Climate Engineering

Climate engineering (sometimes called *geoengineering* or *climate intervention*) is the deliberate modification of the climate. It has been investigated as a possible response to global warming, e.g.

by NASA and the Royal Society. Techniques under research fall generally into the categories solar radiation management and carbon dioxide removal, although various other schemes have been suggested. A study from 2014 investigated the most common climate engineering methods and concluded they are either ineffective or have potentially severe side effects and cannot be stopped without causing rapid climate change.

Discourse About Global Warming

Political Discussion

Article 2 of the UN Framework Convention refers explicitly to "stabilization of greenhouse gas concentrations." To stabilize the atmospheric concentration of CO_2, emissions worldwide would need to be dramatically reduced from their present level.

Most countries in the world are parties to the United Nations Framework Convention on Climate Change (UNFCCC). The ultimate objective of the Convention is to prevent dangerous human interference of the climate system. As stated in the Convention, this requires that GHG concentrations are stabilized in the atmosphere at a level where ecosystems can adapt naturally to climate change, food production is not threatened, and economic development can proceed in a sustainable fashion. The Framework Convention was agreed in 1992, but since then, global emissions have risen.

During negotiations, the G77 (a lobbying group in the United Nations representing 133 developing nations) pushed for a mandate requiring developed countries to "[take] the lead" in reducing their emissions. This was justified on the basis that: the developed world's emissions had contributed most to the cumulation of GHGs in the atmosphere; per-capita emissions (i.e., emissions per head of population) were still relatively low in developing countries; and the emissions of develop-ing countries would grow to meet their development needs.

This mandate was sustained in the Kyoto Protocol to the Framework Convention, which entered into legal effect in 2005. In ratifying the Kyoto Protocol, most developed countries accepted legally binding commitments to limit their emissions. These first-round commitments expired in 2012. United States President George W. Bush rejected the treaty on the basis that "it exempts 80% of the world, including major population centres such as China and India, from compliance, and would cause serious harm to the US economy."

At the 15th UNFCCC Conference of the Parties, held in 2009 at Copenhagen, several UNFCCC Parties produced the Copenhagen Accord. Parties associated with the Accord (140 countries, as of November 2010) aim to limit the future increase in global mean temperature to below °C. The 16th Conference of the Parties (COP16) was held at Cancún in 2010. It produced an agreement, not a binding treaty, that the Parties should take urgent action to reduce greenhouse gas emissions to meet a goal of limiting global warming to °C above pre-industrial temperatures. It also recognized the need to consider strengthening the goal to a global average rise of 5 °C.

Scientific Discussion

There is continuing discussion through published peer-reviewed scientific papers, which are assessed by scientists working in the relevant fields taking part in the Intergovernmental Panel on

Climate Change. The scientific consensus as of 2013 stated in the IPCC Fifth Assessment Report is that it "is extremely likely that human influence has been the dominant cause of the observed warming since the mid-20th century". A 2008 report by the U.S. National Academy of Sciences stated that most scientists by then agreed that observed warming in recent decades was primarily caused by human activities increasing the amount of greenhouse gases in the atmosphere. In 2005 the Royal Society stated that while the overwhelming majority of scientists were in agreement on the main points, some individuals and organizations opposed to the consensus on urgent action needed to reduce greenhouse gas emissions have tried to undermine the science and work of the IPCC. National science academies have called on world leaders for policies to cut global emissions.

In the scientific literature, there is a strong consensus that global surface temperatures have increased in recent decades and that the trend is caused mainly by human-induced emissions of greenhouse gases. No scientific body of national or international standing disagrees with this view.

Discussion by the Public and in Popular Media

The global warming controversy refers to a variety of disputes, substantially more pronounced in the popular media than in the scientific literature, regarding the nature, causes, and consequences of global warming. The disputed issues include the causes of increased global average air temperature, especially since the mid-20th century, whether this warming trend is unprecedented or within normal climatic variations, whether humankind has contributed significantly to it, and whether the increase is completely or partially an artefact of poor measurements. Additional disputes concern estimates of climate sensitivity, predictions of additional warming, and what the consequences of global warming will be.

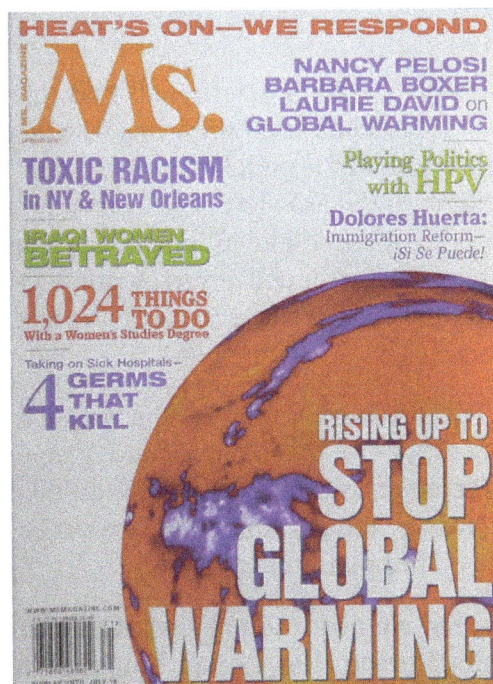

Global warming was the cover story in this 2007 issue of *Ms. magazine*

From 1990 to 1997, right-wing conservative think tanks in the United States mobilized to challenge the legitimacy of global warming as a social problem. They challenged the scientific evidence, argued that global warming will have benefits, and asserted that proposed solutions would do more harm than good. Some people dispute aspects of climate change science. Organizations such as the libertarian Competitive Enterprise Institute, conservative commentators, and some companies such as ExxonMobil have challenged IPCC climate change scenarios, funded scientists who disagree with the scientific consensus, and provided their own projections of the economic cost of stricter controls. On the other hand, some fossil fuel companies have scaled back their efforts in recent years, or even called for policies to reduce global warming. Global oil companies have begun to acknowledge climate change exists and is caused by human activities and the burning of fossil fuels.

Surveys of Public Opinion

The world public, or at least people in economically advanced regions, became broadly aware of the global warming problem in the late 1980s. Polling groups began to track opinions on the subject, at first mainly in the United States. The longest consistent polling, by Gallup in the US, found relatively small deviations of 10% or so from 1998 to 2015 in opinion on the seriousness of global warming, but with increasing polarization between those concerned and those unconcerned.

The first major worldwide poll, conducted by Gallup in 2008-2009 in 127 countries, found that some 62% of people worldwide said they knew about global warming. In the advanced countries of North America, Europe and Japan, 90% or more knew about it (97% in the U.S., 99% in Japan); in less developed countries, especially in Africa, fewer than a quarter knew about it, although many had noticed local weather changes. Among those who knew about global warming, there was a wide variation between nations in belief that the warming was a result of human activities.

By 2010, with 111 countries surveyed, Gallup determined that there was a substantial decrease since 2007–08 in the number of Americans and Europeans who viewed global warming as a serious threat. In the US, just a little over half the population (53%) now viewed it as a serious concern for either themselves or their families; this was 10 points below the 2008 poll (63%). Latin America had the biggest rise in concern: 73% said global warming is a serious threat to their families. This global poll also found that people are more likely to attribute global warming to human activities than to natural causes, except in the US where nearly half (47%) of the population attributed global warming to natural causes.

A March–May 2013 survey by Pew Research Center for the People & the Press polled 39 countries about global threats. According to 54% of those questioned, global warming featured top of the perceived global threats. In a January 2013 survey, Pew found that 69% of Americans say there is solid evidence that the Earth's average temperature has got warmer over the past few decades, up six points since November 2011 and 12 points since 2009.

A 2010 survey of 14 industrialized countries found that skepticism about the danger of global warming was highest in Australia, Norway, New Zealand and the United States, in that order, correlating positively with per capita emissions of carbon dioxide.

Etymology

In the 1950s, research suggested increasing temperatures, and a 1952 newspaper reported "climate change". This phrase next appeared in a November 1957 report in *The Hammond Times* which described Roger Revelle's research into the effects of increasing human-caused CO_2 emissions on the greenhouse effect, "a large scale global warming, with radical climate changes may result". Both phrases were only used occasionally until 1975, when Wallace Smith Broecker published a scientific paper on the topic; "Climatic Change: Are We on the Brink of a Pronounced Global Warming?" The phrase began to come into common use, and in 1976 Mikhail Budyko's statement that "a global warming up has started" was widely reported. Other studies, such as a 1971 MIT report, referred to the human impact as "inadvertent climate modification", but an influential 1979 National Academy of Sciences study headed by Jule Charney followed Broecker in using *global warming* for rising surface temperatures, while describing the wider effects of increased CO_2 as *climate change*.

In 1986 and November 1987, NASA climate scientist James Hansen gave testimony to Congress on global warming. There were increasing heatwaves and drought problems in the summer of 1988, and when Hansen testified in the Senate on 23 June he sparked worldwide interest. He said: "global warming has reached a level such that we can ascribe with a high degree of confidence a cause and effect relationship between the greenhouse effect and the observed warming." Public attention increased over the summer, and *global warming* became the dominant popular term, commonly used both by the press and in public discourse.

In a 2008 NASA article on usage, Erik M. Conway defined *Global warming* as "the increase in Earth's average surface temperature due to rising levels of greenhouse gases", while *Climate change* was "a long-term change in the Earth's climate, or of a region on Earth." As effects such as changing patterns of rainfall and rising sea levels would probably have more impact than temperatures alone, he considered *global climate change* a more scientifically accurate term, and like the Intergovernmental Panel on Climate Change, the NASA website would emphasize this wider context.

Global Warming Controversy

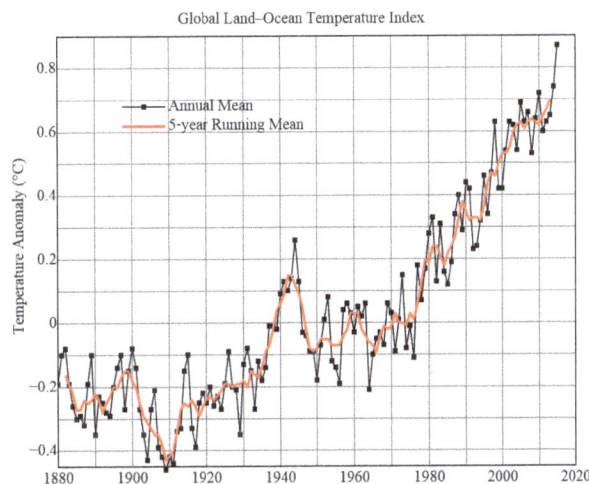

Global mean land-ocean temperature changes from 1880, relative to the 1951–1980 mean. The black line is the annual mean and the red line is the 5-year running mean. Source: NASA GISS.

The global warming controversy concerns the public debate over whether global warming is occurring, how much has occurred in modern times, what has caused it, what its effects will be, whether any action should be taken to curb it, and if so what that action should be. In the scientific literature, there is a strong consensus that global surface temperatures have increased in recent decades and that the trend is caused by human-induced emissions of greenhouse gases. No scientific body of national or international standing disagrees with this view, though a few organizations with members in extractive industries hold non-committal positions. Disputes over the key scientific facts of global warming are more prevalent in the popular media than in the scientific literature, where such issues are treated as resolved, and more prevalent in the United States than globally.

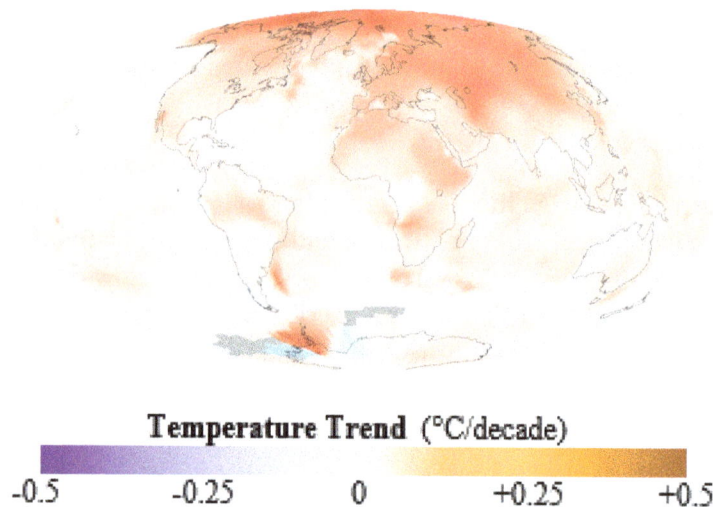

Temperature Trend (°C/decade)

-0.5   -0.25   0   +0.25   +0.5

The map shows the 10-year average (2000–2009) global mean temperature anomaly relative to the 1951–1980 mean. The most extreme warming was in the Arctic. Source: NASA Earth Observatory

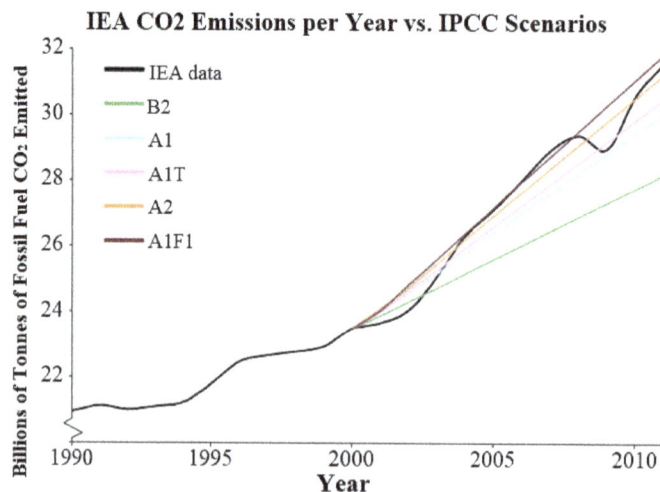

Fossil fuel related CO_2 emissions compared to five of the IPCC's "SRES" emissions scenarios. The dips are related to global recessions. Image source: Skeptical Science.

Political and popular debate concerning the existence and cause of climate change includes the reasons for the increase seen in the instrumental temperature record, whether the warming trend exceeds normal climatic variations, and whether human activities have contributed significantly to it. Scientists have resolved many of these questions decisively in favour of the view that the current warming trend exists and is ongoing, that human activity is the cause, and that it is without precedent in at least 2000 years. Public disputes that also reflect scientific debate include estimates of how responsive the climate system might be to any given level of greenhouse gases (climate sensitivity), how global climate change will play out at local and regional scales, and what the consequences of global warming will be.

Global warming remains an issue of widespread political debate, often split along party political lines, especially in the United States. Many of the issues that are settled within the scientific community, such as human responsibility for global warming, remain the subject of politically or economically motivated attempts to downplay, dismiss or deny them – an ideological phenomenon categorised by academics and scientists as climate change denial. The sources of funding for those involved with climate science – both supporting and opposing mainstream scientific positions – have been questioned by both sides. There are debates about the best policy responses to the science, their cost-effectiveness and their urgency. Climate scientists, especially in the United States, have reported official and oil-industry pressure to censor or suppress their work and hide scientific data, with directives not to discuss the subject in public communications. Legal cases regarding global warming, its effects, and measures to reduce it have reached American courts. The fossil fuels lobby has been identified as overtly or covertly supporting efforts to undermine or discredit the scientific consensus on global warming.

History

Public Opinion

In the United States, the mass media devoted little coverage to global warming until the drought of 1988, and James E. Hansen's testimony to the Senate, which explicitly attributed "the abnormally hot weather plaguing our nation" to global warming.

The British press also changed its coverage at the end of 1988, following a speech by Margaret Thatcher to the Royal Society advocating action against human-induced climate change. According to Anabela Carvalho, an academic analyst, Thatcher's "appropriation" of the risks of climate change to promote nuclear power, in the context of the dismantling of the coal industry following the 1984–1985 miners' strike was one reason for the change in public discourse. At the same time environmental organizations and the political opposition were demanding "solutions that contrasted with the government's". In May 2013 Charles, Prince of Wales took a strong stance criticising both climate change deniers and corporate lobbyists by likening the Earth to a dying patient. "A scientific hypothesis is tested to absolute destruction, but medicine can't wait. If a doctor sees a child with a fever, he can't wait for [endless] tests. He has to act on what is there."

Many European countries took action to reduce greenhouse gas emissions before 1990. West Germany started to take action after the Green Party took seats in Parliament in the 1980s. All countries of the European Union ratified the 1997 Kyoto Protocol. Substantial activity by NGOs took place as well. The United States Energy Information Administration reports that, in the United States, "The 2012 downturn means that emissions are at their lowest level since 1994 and over 12 percent below the recent 2007 peak."

In Europe, the notion of human influence on climate gained wide acceptance more rapidly than in the United States and other countries. A 2009 survey found that Europeans rated climate change as the second most serious problem facing the world, between "poverty, the lack of food and drinking water" and "a major global economic downturn". 87% of Europeans considered climate change to be a very serious or serious problem, while ten per cent did not consider it a serious problem.

In 2007 the BBC announced the cancellation of a planned television special Planet Relief, which would have highlighted the global warming issue and included a mass electrical switch-off. The editor of BBC's Newsnight current affairs show said: "It is absolutely not the BBC's job to save the planet. I think there are a lot of people who think that, but it must be stopped." Author Mark Lynas said "The only reason why this became an issue is that there is a small but vociferous group of extreme right-wing climate 'sceptics' lobbying against taking action, so the BBC is behaving like a coward and refusing to take a more consistent stance."

The authors of the 2010 book *Merchants of Doubt* provide documentation for the assertion that professional deniers have tried to sow seeds of doubt in public opinion in order to halt any meaningful social or political progress to reduce the impact of human carbon emissions. The fact that only half of the American population believe that global warming is caused by human activity could be seen as a victory for these deniers. One of the authors' main arguments is that most prominent scientists who have been voicing opposition to the near-universal consensus are being funded by industries, such as automotive and oil, that stand to lose money by government actions to regulate greenhouse gases.

A compendium of poll results on public perceptions about global warming is below.

Statement	% agree	Year
(US) Global Warming is very/extremely important	49	2006
(International) Climate change is a serious problem.	90	2006
(International) Human activity is a significant cause of climate change.	79	2007
(US) It's necessary to take major steps starting very soon.	59	2007
(US) The Earth is getting warmer because of human activity	49	2009

In 2007 a report on public perceptions in the United Kingdom by Ipsos MORI reported that

- There is widespread recognition that the climate, irrespective of the cause, is changing—88% believe this to be true.

- However, the public is out of step with the scientific community, with 41% believing that climate change is being caused by both human activity and natural processes. 46% believe human activity is the main cause.

- Only a small minority reject anthropogenic climate change, while almost half (44%) are very concerned. However, there remains a large proportion who are not fully persuaded and hold doubts about the extent of the threat.

- There is still a strong appetite among the public for more information, and 63% say they need this to come to a firm view on the issue and what it means for them.

- The public continue to externalize climate change to other people, places and times. It is increasingly perceived as a major global issue with far-reaching consequences for future generations—45% say it is the most serious threat facing the World today and 53% believe it will impact significantly on future generations. However, the issue features less prominently nationally and locally, indeed only 9% believe climate change will have a significant impact upon them personally.

The Canadian science broadcaster and environmental activist, David Suzuki, reports that focus groups organized by the David Suzuki Foundation in 2006 showed that the public has a poor understanding of the science behind global warming. This is despite publicity through different means, including the films *An Inconvenient Truth* and *The 11th Hour*.

An example of the poor understanding is public confusion between global warming and ozone depletion or other environmental problems.

A 15-nation poll conducted in 2006 by Pew Global found that there "is a substantial gap in concern over global warming—roughly two-thirds of Japanese (66%) and Indians (65%) say they personally worry a great deal about global warming. Roughly half of the populations of Spain (51%) and France (46%) also express great concern over global warming, based on those who have heard about the issue. But there is no evidence of alarm over global warming in either the United States or China—the two largest producers of greenhouse gases. Just 19% of Americans and 20% of the Chinese who have heard of the issue say they worry a lot about global warming—the lowest percentages in the 15 countries surveyed. Moreover, nearly half of Americans (47%) and somewhat fewer Chinese (37%) express little or no concern about the problem."

A 47-nation poll by Pew Global Attitudes conducted in 2007 found, "Substantial majorities 25 of 37 countries say global warming is a 'very serious' problem."

There are differences between the opinion of scientists and that of the general public. A 2009 poll in the US by Pew Research Center found "[w]hile 84% of scientists say the earth is getting warmer because of human activity such as burning fossil fuels, just 49% of the public agrees". A 2010 poll in the UK for the BBC showed "Climate scepticism on the rise". Robert Watson found this "very disappointing" and said "We need the public to understand that climate change is serious so they will change their habits and help us move towards a low carbon economy."

A 2012 Canadian poll found that 32% of Canadians said they believe climate change is happening because of human activity, while 54% said they believe it's because of human activity and partially due to natural climate variation. 9% believe climate change is occurring due to natural climate variation, and only 2% said they don't believe climate change is occurring at all.

Related Controversies

Many of the critics of the consensus view on global warming have disagreed, in whole or part, with the scientific consensus regarding other issues, particularly those relating to environmental risks, such as ozone depletion, DDT, and passive smoking. Chris Mooney, author of *The Republican War on Science*, has argued that the appearance of overlapping groups of skeptical scientists, commentators and think tanks in seemingly unrelated controversies results from an organized attempt to replace scientific analysis with political ideology. Mooney says that the promotion of doubt regarding issues that are politically, but not scientifically, controversial became increasingly prevalent under the George W. Bush administration, which, he says, regularly distorted and/or suppressed scientific research to further its own political aims. This is also the subject of a 2004 book by environmental lawyer Robert F. Kennedy, Jr. titled *Crimes Against Nature: How George W. Bush and Corporate Pals are Plundering the Country and Hijacking Our Democracy* (ISBN 978-0060746872). Another book on this topic is *The Assault on Reason* by former Vice President of the United States Al Gore. Earlier instances of this trend are also covered in the book *The Heat Is On* by Ross Gelbspan.

Some critics of the scientific consensus on global warming have argued that these issues should not be linked and that reference to them constitutes an unjustified ad hominem attack. Political scientist Roger Pielke, Jr., responding to Mooney, has argued that science is inevitably intertwined with politics.

In 2015, according to *The New York Times* and others, oil companies knew that burning oil and gas could cause global warming since the 1970s but, nonetheless, funded deniers for years.

Mainstream Scientific Position, and Challenges to it

The finding that the climate has warmed in recent decades and that human activities are producing global climate change has been endorsed by every national science academy that has issued a statement on climate change, including the science academies of all of the major industrialized countries.

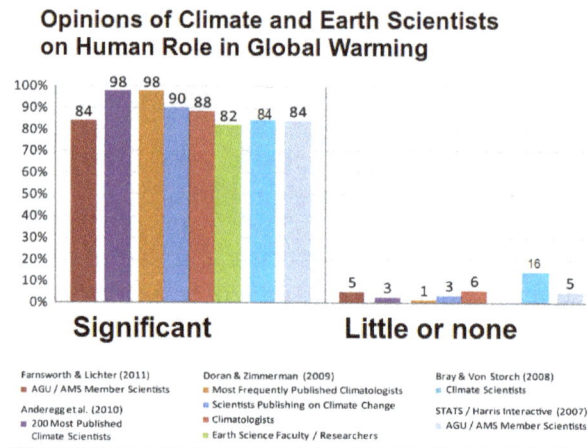

Summary of opinions from climate and earth scientists regarding climate change.

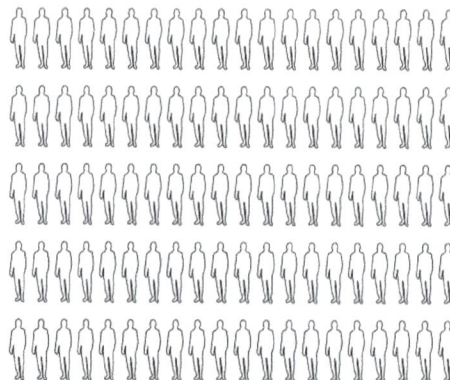

Just over 97% of climate researchers say humans are causing most global warming.

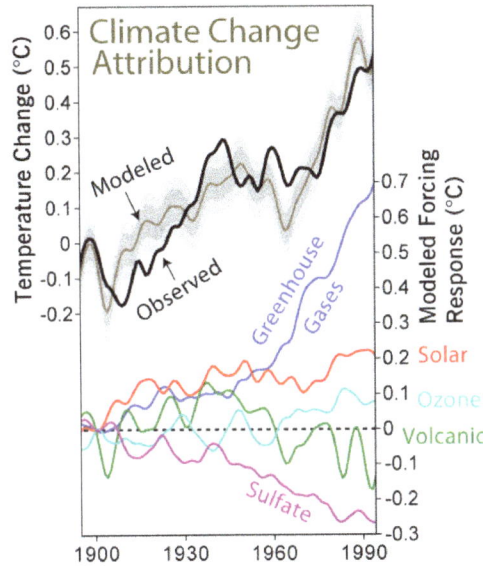

Reproduction of the temperature record using historical forcings

Attribution of recent climate change discusses how global warming is attributed to anthropogenic greenhouse gases (GHGs).

Scientific Consensus

Scientific consensus is normally achieved through communication at conferences, publication in the scientific literature, replication (reproducible results by others), and peer review. In the case of global warming, many governmental reports, the media in many countries, and environmental groups, have stated that there is virtually unanimous scientific agreement that human-caused global warming is real and poses a serious concern. According to the United States National Research Council,

[T]here is a strong, credible body of evidence, based on multiple lines of research, documenting that climate is changing and that these changes are in large part caused by human activities. While much remains to be learned, the core phenomenon, scientific questions, and hypotheses have been examined thoroughly and have stood firm in the face of serious scientific debate and careful evaluation of alternative explanations. * * * Some scientific conclusions or theories have been so thoroughly examined and tested, and supported by so many independent observations and results, that their likelihood of subsequently being found to be wrong is vanishingly small. Such conclusions and theories are then regarded as settled facts. This is the case for the conclusions that the Earth system is warming and that much of this warming is very likely due to human activities.

Among opponents of the mainstream scientific assessment, some say that while there is agreement that humans do have an effect on climate, there is no universal agreement about the quantitative magnitude of anthropogenic global warming (AGW) relative to natural forcings and its harm to benefit ratio. Other opponents assert that some kind of ill-defined "consensus argument" is being used, and then dismiss this by arguing that science is based on facts rather than consensus. Some highlight the dangers of focusing on only one viewpoint in the context of what they say is unsettled science, or point out that science is based on facts and not on opinion polls or consensus.

Dennis T. Avery, a food policy analyst at the Hudson Institute, wrote an article titled "500 Scientists Whose Research Contradicts Man-Made Global Warming Scares" published in 2007 by The Heartland Institute. The list was immediately called into question for misunderstanding and distorting the conclusions of many of the named studies and citing outdated, flawed studies that had long been abandoned. Many of the scientists included in the list demanded their names be removed. The Heartland Institute refused these requests, stating that the scientists "have no right—legally or ethically—to demand that their names be removed from a bibliography composed by researchers with whom they disagree" despite the aforementioned falsification and refutation of much of the list.

A 2010 paper in the Proceedings of the National Academy of Sciences analysed "1,372 climate researchers and their publication and citation data to show that (i) 97–98% of the climate researchers most actively publishing in the field support the tenets of ACC outlined by the Intergovernmental Panel on Climate Change, and (ii) the relative climate expertise and scientific prominence of the researchers unconvinced of ACC are substantially below that of the convinced researchers". Judith Curry has said "This is a completely unconvincing analysis", whereas Naomi Oreskes said that the paper shows "the vast majority of working [climate] research scientists are in agreement [on climate change]... Those who don't agree, are, unfortunately—and this is hard to say without sounding elitist—mostly either not actually climate researchers or not very productive researchers." Jim Prall, one of the coauthors of the study, acknowledged "it would be helpful to have lukewarm [as] a third category."

A 2013 study published in the peer-reviewed journal *Environmental Research Letters* analyzed 11,944 abstracts from papers published in the peer-reviewed scientific literature between 1991 and 2011, identified by searching the ISI Web of Science citation index engine for the text strings "global climate change" or "global warming". The authors found that 3974 of the abstracts expressed a position on anthropogenic global warming, and that 97.1% of those endorsed the consensus that humans are causing global warming. The authors found that of the 11,944 abstracts, 3896 endorsed that consensus, 7930 took no position on it, 78 rejected the consensus, and 40 expressed uncertainty about it.

In 2014 a letter from 52 leading skeptics was published by the Committee for Skeptical Inquiry supporting the scientific consensus and asking the media to stop referring to deniers as "skeptics." The letter clarified the skeptical opinion on climate and denial: "As scientific skeptics, we are well aware of political efforts to undermine climate science by those who deny reality but do not engage in scientific research or consider evidence that their deeply held opinions are wrong. The most appropriate word to describe the behavior of those individuals is 'denial'. Not all individuals who call themselves climate change skeptics are deniers. But virtually all deniers have falsely branded themselves as skeptics. By perpetrating this misnomer, journalists have granted undeserved credibility to those who reject science and scientific inquiry."

Authority of the IPCC

The "standard" view of climate change has come to be defined by the reports of the IPCC, which is supported by many other science academies and scientific organizations. In 2001, sixteen of the world's national science academies made a joint statement on climate change, and gave their support for the IPCC.

Opponents have generally attacked either the IPCC's processes, people or the Synthesis and Executive summaries; the full reports attract less attention. Some of the controversy and criticism has originated from experts invited by the IPCC to submit reports or serve on its panels. For example, Richard Lindzen has publicly dissented from IPCC positions.

Christopher Landsea, a hurricane researcher, said of "the part of the IPCC to which my expertise is relevant" that "I personally cannot in good faith continue to contribute to a process that I view as both being motivated by pre-conceived agendas and being scientifically unsound," because of comments made at a press conference by Kevin Trenberth of which Landsea disapproved. Trenberth said "Landsea's comments were not correct"; the IPCC replied "individual scientists can do what they wish in their own rights, as long as they are not saying anything on behalf of the IPCC" and offered to include Landsea in the review phase of the AR4. Roger Pielke, Jr. commented "Both Landsea and Trenberth can and should feel vindicated... the IPCC accurately reported the state of scientific understandings of tropical cyclones and climate change in its recent summary for policy makers."

In 2005, the House of Lords Economics Committee wrote, "We have some concerns about the objectivity of the IPCC process, with some of its emissions scenarios and summary documentation apparently influenced by political considerations." It doubted the high emission scenarios and said that the IPCC had "played-down" what the committee called "some positive aspects of global warming". The main statements of the House of Lords Economics Committee were rejected in the response made by the United Kingdom government and by the Stern Review.

Speaking to the difficulty of establishing scientific consensus on the precise extent of human action on climate change, John Christy, a contributing author, wrote:

Contributing authors essentially are asked to contribute a little text at the beginning and to review the first two drafts. We have no control over editing decisions. Even less influence is granted the 2,000 or so reviewers. Thus, to say that 800 contributing authors or 2,000 reviewers reached consensus on anything describes a situation that is not reality.

On 10 December 2008, a report was released by the U.S. Senate Committee on Environment and Public Works Minority members, under the leadership of the Senate's most vocal global warming skeptic Jim Inhofe. The timing of the report coincided with the UN global warming conference in Poznań, Poland. It says it summarizes scientific dissent from the IPCC. Many of its statements about the numbers of individuals listed in the report, whether they are actually scientists, and whether they support the positions attributed to them, have been disputed.

While some critics have argued that the IPCC overstates likely global warming, others have made the opposite criticism. David Biello, writing in the Scientific American, argues that, because of the need to secure consensus among governmental representatives, the IPCC reports give conservative estimates of the likely extent and effects of global warming. *Science* editor Brooks Hanson states in a 2010 editorial: "The IPCC reports have underestimated the pace of climate change while overestimating societies' abilities to curb greenhouse gas emissions." Climate scientist James E. Hansen argues that the IPCC's conservativeness seriously underestimates the risk of sea-level rise on the order of meters—enough to inundate many low-lying areas, such as the southern third of Florida. Roger A. Pielke Sr. has also stated "Humans are significantly altering the global climate, but in a variety of diverse ways beyond the radiative effect of carbon dioxide. The IPCC assessments have been too conservative in recognizing the importance of these human climate forcings as they alter regional and global climate."

Henderson-Sellers has collected comments from IPCC authors in a 2007 workshop revealing a number of concerns. She concluded, "Climate change research entered a new and different regime with the publication of the IPCC Fourth Assessment Report. There is no longer any question about "whether" human activities are changing the climate; instead research must tackle the urgent questions of: "how fast?"; "with what impacts?'; and "what responses are needed?""

Greenhouse Gases

Attribution of recent climate change discusses the evidence for recent global warming. Correlation of CO_2 and temperature is not part of this evidence. Nonetheless, one argument against global warming says that rising levels of carbon dioxide (CO_2) and other greenhouse gases (GHGs) do not correlate with global warming.

- Studies of the Vostok ice core show that at the "beginning of the deglaciations, the CO_2 increase either was in phase or lagged by less than ~1000 years with respect to the Antarctic temperature, whereas it clearly lagged behind the temperature at the onset of the glaciations". Recent warming is followed by carbon dioxide levels with only a 5 months delay. The time lag has been used to argue that the current rise in CO_2 is a *result* of warming and not a cause. While it is generally agreed that variations before the industrial age are mostly timed by astronomical forcing, a main part of current warming is found to be timed by anthropogenic releases of CO_2, having a much closer time relation not observed in the past (thus returning the argument to the importance of human CO_2 emissions). Analysis of carbon isotopes in atmospheric CO_2 shows that the recent observed CO_2 increase cannot have come from the oceans, volcanoes, or the biosphere, and thus is not a response to rising temperatures as would be required if the same processes creating past lags were active now.

- Carbon dioxide accounts for about 390 parts per million by volume (ppm) of the Earth's atmosphere, increasing from 284 ppm in the 1830s to 387 ppm in 2009. Carbon dioxide contributes between 9 and 26% of the natural greenhouse effect.

- In the Ordovician period of the Paleozoic era (about 450 million years ago), the Earth had an atmospheric CO_2 concentration estimated at 4400ppm (or 0.44% of the atmosphere), while also having evidence of some glaciation. Modeling work has shown that it is possible for local areas at elevations greater than 300–500 meters to contain year-round snow cover even with high atmospheric CO_2 concentrations. A 2006 study suggests that the elevated CO_2 levels and the glaciation are not synchronous, but rather that weathering associated with the uplift and erosion of the Appalachian Mountains greatly reduced atmospheric greenhouse gas concentrations and permitted the observed glaciation.

As noted above, climate models are only able to simulate the temperature record of the past century when GHG forcing is included, being consistent with the findings of the IPCC which has stated that: "Greenhouse gas forcing, largely the result of human activities, has very likely caused most of the observed global warming over the last 50 years"

The "standard" set of scenarios for future atmospheric greenhouse gases are the IPCC SRES scenarios. The purpose of the range of scenarios is not to predict what exact course the future of emissions will take, but what it may take under a range of possible population, economic and societal trends. Climate models can be run using any of the scenarios as inputs to illustrate the different outcomes for climate change. No one scenario is officially preferred, but in practice the "A1b" scenario roughly corresponding to 1%/year growth in atmospheric CO_2 is often used for modelling studies.

There is debate about the various scenarios for fossil fuel consumption. Global warming skeptic Fred Singer stated "some good experts believe" that atmospheric CO_2 concentration will not double since economies are becoming less reliant on carbon.

However, the Stern report, like many other reports, notes the past correlation between CO_2 emissions and economic growth and then extrapolates using a "business as usual" scenario to predict GDP growth and hence CO_2 levels, concluding that:

Increasing scarcity of fossil fuels alone will not stop emissions growth in time. The stocks of hydrocarbons that are profitable to extract are more than enough to take the world to levels of CO_2 well beyond 750 ppm with very dangerous consequences for climate change impacts.

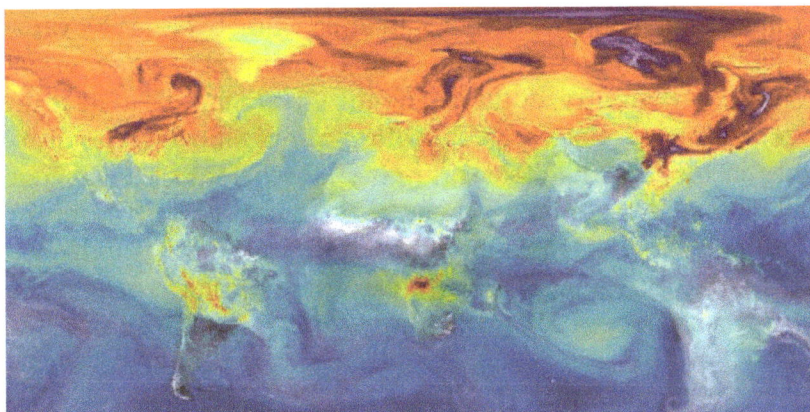

CO_2 in Earth's atmosphere if *half* of global-warming emissions are *not* absorbed.
(NASA computer simulation).

According to a 2006 paper from Lawrence Livermore National Laboratory, "the earth would warm by 8 degrees Celsius (14.4 degrees Fahrenheit) if humans use the entire planet's available fossil fuels by the year 2300."

On 12 November 2015, NASA scientists reported that human-made carbon dioxide (CO_2) continues to increase above levels not seen in hundreds of thousands of years: currently, about half of the carbon dioxide released from the burning of fossil fuels remains in the atmosphere and is not absorbed by vegetation and the oceans.

Solar Variation

Scientists opposing the mainstream scientific assessment of global warming express varied opinions concerning the cause of global warming. Some say only that it has not yet been ascertained whether humans are the primary cause of global warming; others attribute global warming to natural variation; ocean currents; increased solar activity or cosmic rays. The consensus position is that solar radiation may have increased by 0.12 W/m² since 1750, compared to 1.6 W/m² for the net anthropogenic forcing. The TAR said, "The combined change in radiative forcing of the two major natural factors (solar variation and volcanic aerosols) is estimated to be negative for the past two, and possibly the past four, decades." The AR4 makes no direct assertions on the recent role of solar forcing, but the previous statement is consistent with the AR4's figure 4.

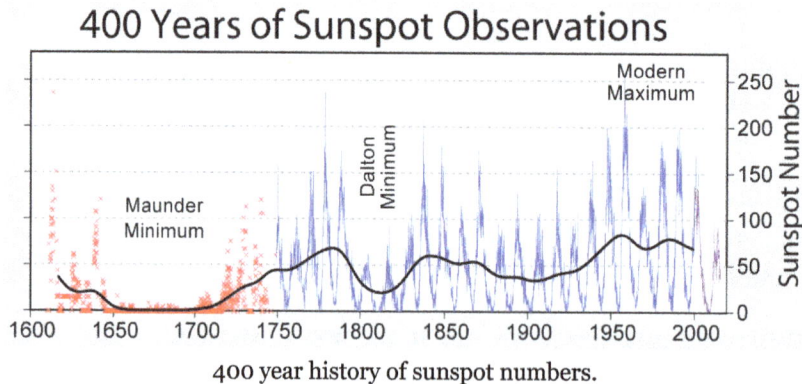

400 year history of sunspot numbers.

A few studies say that the present level of solar activity is historically high as determined by sunspot activity and other factors. Solar activity could affect climate either by variation in the Sun's output or, more speculatively, by an indirect effect on the amount of cloud formation. Solanki and co-workers suggest that solar activity for the last 60 to 70 years may be at its highest level in 8,000 years; Muscheler *et al.* disagree, suggesting that other comparably high levels of activity have occurred several times in the last few thousand years. Muscheler *et al.* concluded "solar activity reconstructions tell us that only a minor fraction of the recent global warming can be explained by the variable Sun." Solanki *et al.* concluded "that solar variability is unlikely to have been the dominant cause of the strong warming during the past three decades", and "at the most 30% of the strong warming since then can be of solar origin".

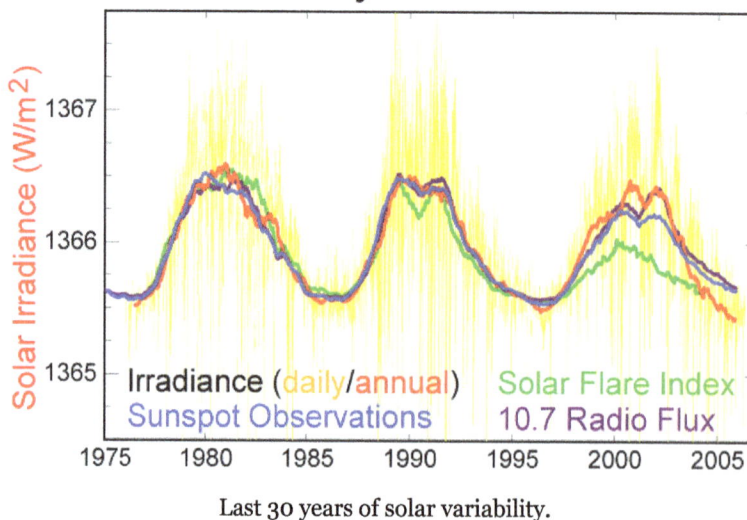

Last 30 years of solar variability.

Another point of controversy is the correlation of temperature with solar variation.

Mike Lockwood and Claus Fröhlich reject the statement that the warming observed in the global mean surface temperature record since about 1850 is the result of solar variations. Lockwood and Fröhlich conclude, "the observed rapid rise in global mean temperatures seen after 1985 cannot be ascribed to solar variability, whichever of the mechanisms is invoked and no matter how much the solar variation is amplified."

Aerosols Forcing

The hiatus in warming from the 1940s to 1960s is generally attributed to cooling effect of sulphate aerosols. More recently, this forcing has (relatively) declined, which may have enhanced warming, though the effect is regionally varying. Another example of this is in Ruckstuhl's paper who found a 60% reduction in aerosol concentrations over Europe causing solar brightening:

[...] the direct aerosol effect had an approximately five times larger impact on climate forcing than the indirect aerosol and other cloud effects. The overall aerosol and cloud induced surface climate forcing is ~ 1 W m^{-2} decade^{-1} and has most probably strongly contributed to the recent rapid warming in Europe.

Analysis of Temperature Records

Instrumental Record of Surface Temperature

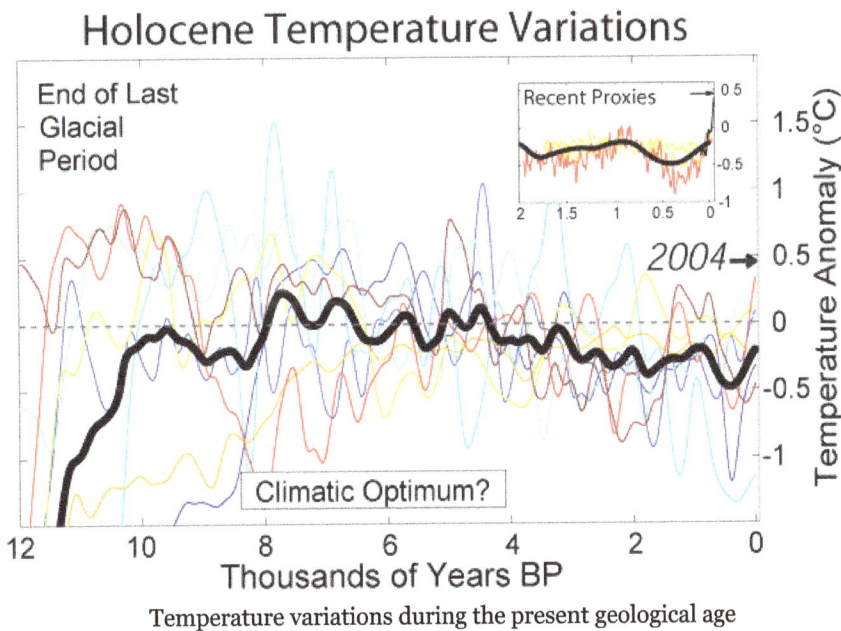

Temperature variations during the present geological age

There have been attempts to raise public controversy over the accuracy of the instrumental temperature record on the basis of the urban heat island effect, the quality of the surface station network, and assertions that there have been unwarranted adjustments to the temperature record.

Weather stations that are used to compute global temperature records are not evenly distributed over the planet. There were a small number of weather stations in the 1850s, and the number didn't reach the current 3000+ until the 1951 to 1990 period

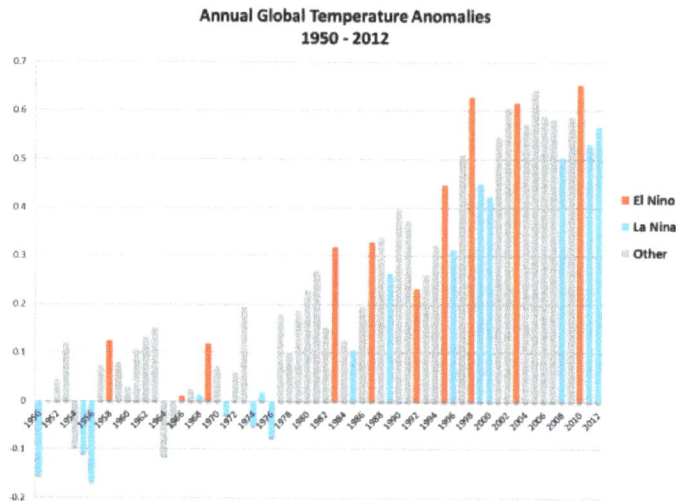

Annual Global Temperature Anomalies
1950 - 2012

NOAA graph of Global Annual Temperature Anomalies 1950–2012

The 2001 IPCC Third Assessment Report (TAR) acknowledged that the urban heat island is an important *local* effect, but cited analyses of historical data indicating that the effect of the urban heat island on the *global* temperature trend is no more than 0.05 °C (0.09 °F) degrees through 1990. Peterson (2003) found no difference between the warming observed in urban and rural areas.

Parker (2006) found that there was no difference in warming between calm and windy nights. Since the urban heat island effect is strongest for calm nights and is weak or absent on windy nights, this was taken as evidence that global temperature trends are not significantly contaminated by urban effects. Pielke and Matsui published a paper disagreeing with Parker's conclusions.

In 2005, Roger A. Pielke and Stephen McIntyre criticized the US instrumental temperature record and adjustments to it, and Pielke and others criticized the poor quality siting of a number of weather stations in the United States. In 2007, Anthony Watts began a volunteer effort to photographically document the siting quality of these stations. The *Journal of Geophysical Research – Atmospheres* subsequently published a study by Menne et al. which examined the record of stations picked out by Watts' Surfacestations.org and found that, if anything, the poorly sited stations showed a slight cool bias rather than the warm bias which Watts had anticipated.

The Berkeley Earth Surface Temperature group carried out an independent assessment of land temperature records, which examined issues raised by skeptics, such as the urban heat island effect, poor station quality, and the risk of data selection bias. The preliminary results, made public in October 2011, found that these factors had not biased the results obtained by NOAA, the Hadley Centre together with the Climatic Research Unit (HadCRUT) and NASA's GISS in earlier studies. The group also confirmed that over the past 50 years the land surface warmed by 0.911 °C, and their results closely matched those obtained from these earlier studies. The four papers they had produced had been submitted for peer review.

Instrumental Record of Tropospheric Temperature

General circulation models and basic physical considerations predict that in the tropics the temperature of the troposphere should increase more rapidly than the temperature of the surface.

A 2006 report to the U.S. Climate Change Science Program noted that models and observations agreed on this amplification for monthly and interannual time scales but not for decadal time scales in most observed data sets. Improved measurement and analysis techniques have reconciled this discrepancy: corrected buoy and satellite surface temperatures are slightly cooler and corrected satellite and radiosonde measurements of the tropical troposphere are slightly warmer. Satellite temperature measurements show that tropospheric temperatures are increasing with "rates similar to those of the surface temperature", leading the IPCC to conclude that this discrepancy is reconciled.

Geologic Temperature Records

Before humans learned to record the temperature of earth's climate system, various biological and geological processes left clues to past climate conditions. The analysis of these clues is the focus of the science of paleoclimatology. The field still has a variety of uncertainties.

Antarctica Cooling

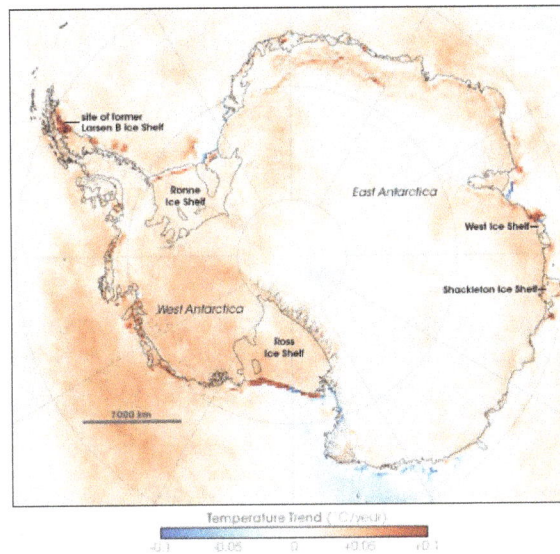

Antarctic Skin (the roughly top millimeter of land, sea, snow, or ice) Temperature Trends between 1981 and 2007, based on thermal infrared observations made by a series of NOAA satellite sensors; note that they do not necessarily reflect air temperature trends.

There has been a public dispute regarding the apparent contradiction in the observed behavior of Antarctica, as opposed to the global rise in temperatures measured elsewhere in the world. This became part of the public debate in the global warming controversy, particularly between advocacy groups of both sides in the public arena, as well as the popular media.

In contrast to the popular press, there is no evidence of a corresponding controversy in the scientific community. Observations unambiguously show the Antarctic Peninsula to be warming. The trends elsewhere show both warming and cooling but are smaller and dependent on season and the timespan over which the trend is computed. A study released in 2009, combined historical weather station data with satellite measurements to deduce past temperatures over large regions of the continent, and these temperatures indicate an overall warming trend. One of the paper's authors stated "We now see warming is taking place on all seven of the earth's continents in accord with what models predict as a response to greenhouse gases." According to 2011 paper by Ding, et al., "The Pacific sector of Antarctica, including both the Antarctic Peninsula and continental West Antarctica, has experienced substantial warming in the past 30 years."

This controversy began with the misinterpretation of the results of a 2002 paper by Doran et al., which found "Although previous reports suggest slight recent continental warming, our spatial analysis of Antarctic meteorological data demonstrates a net cooling on the Antarctic continent between 1966 and 2000, particularly during summer and autumn." Later the controversy was popularized by Michael Crichton's 2004 fiction novel State of Fear, who advocated skepticism in global warming. This novel has a docudrama plot based upon the idea that there is a deliberately alarmist conspiracy behind global warming activism. One of the characters argues "data show that one relatively small area called the Antarctic Peninsula is melting and calving huge icebergs... but the continent as a whole is getting colder, and the ice is getting thicker." As a basis for this plot twist, Crichton cited the peer reviewed scientific article by Doran, et al. Peter Doran, the lead author of the paper cited by Crichton, stated "... our results have been misused as 'evidence' against global warming by Crichton in his novel 'State of Fear'... 'Our study did find that 58 percent of Antarctica cooled from 1966 to 2000. But during that period, the rest of the continent was warming. And climate models created since our paper was published have suggested a link between the lack of significant warming in Antarctica and the ozone hole over that continent."

Climate Sensitivity

As defined by the IPCC, climate sensitivity is the "equilibrium temperature rise that would occur for a doubling of CO_2 concentration above pre-industrial levels". In its 2007 Fourth Assessment Report, IPCC said that climate sensitivity is "likely to be in the range 2 to 4.5 °C with a best estimate of about 3 °C".

Using a combination of surface temperature history and ocean heat content, Stephen E. Schwartz has proposed an estimate of climate sensitivity of 1.9 ± 1.0 K for doubled CO_2., revised upwards from 1.1 ± 0.5 K. Grant Foster, James Annan, Gavin Schmidt, and Michael E. Mann argue that there are errors in both versions of Schwartz's analysis. Petr Chylek and co-authors have also proposed low climate sensitivity to doubled CO_2, estimated to be 1.6 K ± 0.4 K.

In January 2013 widespread publicity was given to work led by Terje Berntsen of the University of Oslo, Julia Hargreaves of the Research Institute for Global Change in Yokohama, and Nic Lewis, an independent climate scientist, which reportedly found lower climate sensitivities than IPCC estimates and the suggestion that there is a 90% probability that doubling CO_2 emissions will increase temperatures by lower values than those estimated by the climate models used by the IPCC was featured in news outlets including The Economist. This premature announcement came from a preliminary news release about a study which had not yet been peer reviewed. The Center for International Climate and Environmental Research, Oslo (CICERO) issued a statement that they were involved with the relevant research project, and the news story was based on a report submitted to the research council which included both published and unpublished material. The highly publicised figures came from work still undergoing peer review, and CICERO would wait until they had been published in a journal before disseminating the results.

Infrared Iris Hypothesis

In 2001, Richard Lindzen proposed a system of compensating meteorological processes involving clouds that tend to stabilize climate change; he tagged this the "Iris hypothesis, or "Infrared Iris". This work has been discussed in a number of papers

Roy Spencer *et al.* suggested "a net reduction in radiative input into the ocean-atmosphere system" in tropical intraseasonal oscillations "may potentially support" the idea of an "Iris" effect, although they point out that their work is concerned with much shorter time scales.

Other analyses have found that the iris effect is a *positive* feedback rather than the negative feedback proposed by Lindzen.

Internal Radiative Forcing

Roy Spencer hypothesized in 2008 that there is an "internal radiative forcing" affecting climate variability,

[...] mixing up of cause and effect when observing natural climate variability can lead to the mistaken conclusion that the climate system is more sensitive to greenhouse gas emissions than it really is. [...] it provides a quantitative mechanism for the (minority) view that global warming is mostly a manifestation of natural internal climate variability.

[...] low frequency, internal radiative forcing amounting to little more than 1 W/m², assumed to be proportional to a weighted average of the southern oscillation and Pacific decadal oscillation indices since 1900, produces ocean temperature behavior similar to that observed: warming from 1900 to 1940, then slight cooling through the 1970s, then resumed warming up to the present, as well as 70% of the observed centennial temperature trend.

Temperature Projections

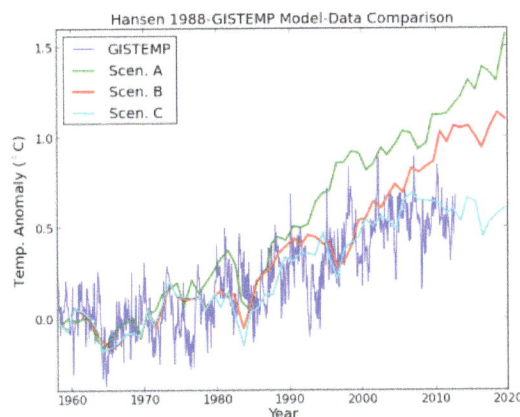

James Hansen's 1988 climate model projections compared with the GISS measured temperature record.

James Hansen's 1984 climate model projections versus observed temperatures are updated each year by Dr Mikako Sato of Columbia University. The RealClimate website provides an annual update comparing both Hansen's 1988 model projections and the IPCC Fourth Assessment Report (AR4) climate model projections with observed temperatures recorded by GISS and HadCRUT. The measured temperatures show continuing global warming.

Conventional projections of future temperature rises depend on estimates of future anthropogenic GHG emissions, those positive and negative climate change feedbacks that have so far been incorporated into the models, and the climate sensitivity. Models referenced by the Intergov-ernmental Panel on Climate Change (IPCC) predict that global temperatures are likely to increase by 1.1 to 6.4 °C (2.0 to 11.5 °F) between 1990 and 2100. Others have proposed that temperature increases may be higher than IPCC estimates. One theory is that the climate may reach a "tipping point" where positive feedback effects lead to runaway global warming; such feedbacks include decreased reflection of solar radiation as sea ice melts, exposing darker seawater, and the potential release of large volumes of methane from thawing permafrost. In 1959 Dr. Bert Bolin, in a speech to the National Academy of Sciences, predicted that by the year 2000 there would be a 25% in-crease in carbon dioxide in the atmosphere compared to the levels in 1859. The actual increase by 2000 was about 29%.

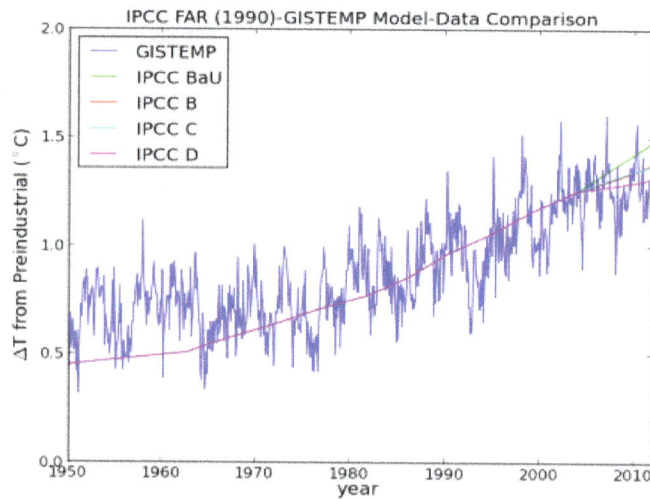

IPCC AR4 projections compared to the GISS temperature record.

David Orrell or Henk Tennekes say that climate change cannot be accurately predicted. Orrell says that the range of future increase in temperature suggested by the IPCC rather represents a social consensus in the climate community, but adds "we are having a dangerous effect on the climate".

A 2007 study by David Douglass and coworkers concluded that the 22 most commonly used global climate models used by the IPCC were unable to accurately predict accelerated warming in the troposphere although they did match actual surface warming, concluding "projections of future climate based on these models should be viewed with much caution". This result went against a similar study of 19 models which found that discrepancies between model predictions and actual temperature were likely due to measurement errors.

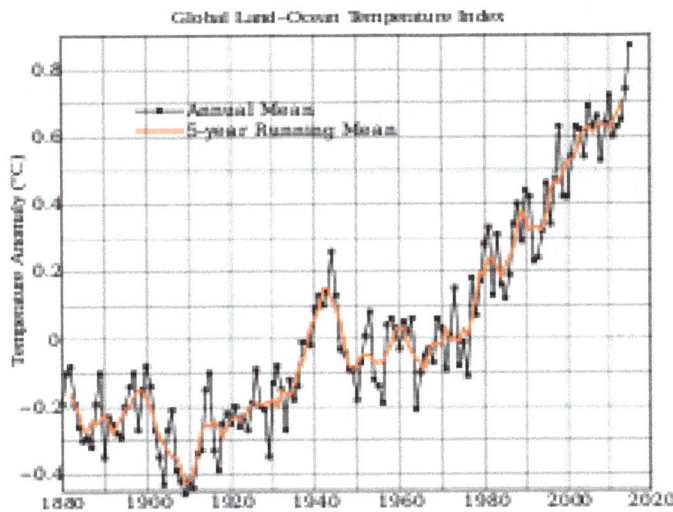

Global mean land-ocean temperature changes from 1880, relative to the 1951–1980 mean. Source: NASA GISS.

In a NASA report published in January 2013, Hansen and Sato noted "the 5-year mean global temperature has been flat for a decade, which we interpret as a combination of natural variability and a slowdown in the growth rate of the net climate forcing." Ed Hawkins, of the University of Reading, stated that the "surface temperatures since 2005 are already at the low end of the range of projections derived from 20 climate models. If they remain flat, they will fall outside the models' range within a few years." Using the long-term temperature trends for the earth scientists and statisticians conclude that it continues to warm through time.

Forecasts Confidence

The IPCC states it has increased confidence in forecasts coming from General Circulation Models or GCMs. Chapter 8 of AR4 reads:

There is considerable confidence that climate models provide credible quantitative estimates of future climate change, particularly at continental scales and above. This confidence comes from the foundation of the models in accepted physical principles and from their ability to reproduce observed features of current climate and past climate changes. Confidence in model estimates is higher for some climate variables (e.g., temperature) than for others (e.g., precipitation). Over several decades of development, models have consistently provided a robust and unambiguous picture of significant climate warming in response to increasing greenhouse gases.

Certain scientists, skeptics and otherwise, believe this confidence in the models' ability to predict future climate is not earned.

Arctic Sea Ice Decline

Following the (then) record low of the arctic sea ice extend in 2007, Mark Serreze, the director of US National Snow and Ice Data Center, stated "If you asked me a couple of years ago when the Arctic could lose all of its ice then I would have said 2100, or 2070 maybe. But now I think that 2030 is a reasonable estimate." In 2012, during another record low, Peter Wadhams of Cambridge University predicted a possible final collapse of Arctic sea ice in summer months around 2016.

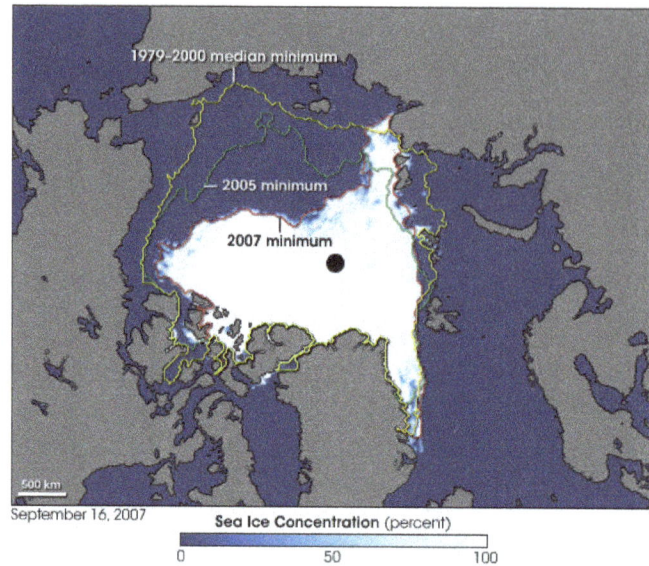

Arctic Sea ice as of 2007 compared to 2005 and also compared to 1979–2000 average

Antarctic and Arctic sea ice extent are available on a daily basis from the National Snow & Ice Data Center.

Data Archiving and Sharing

Scientific journals and funding agencies generally require authors of peer-reviewed research to provide information on archives of data and share sufficient data and methods necessary for a scientific expert on the topic to reproduce the work.

In political controversy over the 1998 and 1999 historic temperature reconstructions widely publicised as the "hockey stick graphs", Mann, Bradley and Hughes as authors of the studies were sent letters on 23 June 2005 from Rep. Joe Barton, chairman of the House Committee on Energy and Commerce and Ed Whitfield, Chairman of the Subcommittee on Oversight and Investigations, demanding full records on the research. The letters told the scientist to provide not just data and methods, but also personal information about their finances and careers, information about grants provided to the institutions they had worked for, and the exact computer codes used to generate their results.

Sherwood Boehlert, chairman of the House Science Committee, told his fellow Republican Joe Barton it was a "misguided and illegitimate investigation" seemingly intended to "intimidate scientists rather than to learn from them, and to substitute congressional political review for scientific review". The U.S. National Academy of Sciences (NAS) president Ralph J. Cicerone wrote to Barton proposing that the NAS should appoint an independent panel to investigate. Barton dismissed this offer.

On 15 July, Mann wrote giving his detailed response to Barton and Whitfield. He emphasised that the full data and necessary methods information was already publicly available in full accordance with National Science Foundation (NSF) requirements, so that other scientists had been able to reproduce their work. NSF policy was that computer codes are considered the intellectual property of researchers and are not subject to disclosure, but notwithstanding these property rights, the program used to generate the original MBH98 temperature reconstructions had been made available at the Mann et al. public FTP site.

Many scientists protested about Barton's demands. Alan I. Leshner wrote to him on behalf of the American Association for the Advancement of Science stating that the letters gave "the impression of a search for some basis on which to discredit these particular scientists and findings, rather than a search for understanding", He stated that Mann, Bradley and Hughes had given out their full data and descriptions of methods. A *Washington Post* editorial on 23 July which described the investigation as harassment quoted Bradley as saying it was "intrusive, far-reaching and intimidating", and Alan I. Leshner of the AAAS describing it as unprecedented in the 22 years he had been a government scientist; he thought it could "have a chilling effect on the willingness of people to work in areas that are politically relevant". Congressman Boehlert said the investigation was as "at best foolhardy" with the tone of the letters showing the committee's inexperience in relation to science.

Barton was given support by global warming sceptic Myron Ebell of the Competitive Enterprise Institute, who said "We've always wanted to get the science on trial ... we would like to figure out a way to get this into a court of law," and "this could work". In his *Junk Science* column on Fox News, Steven Milloy said Barton's inquiry was reasonable. In September 2005 David Legates alleged in a newspaper op-ed that the issue showed climate scientists not abiding by data access requirements and suggested that legislators might ultimately take action to enforce them.

Political Questions

In the U.S. global warming is often a partisan political issue. Republicans tend to oppose action against a threat that they regard as unproven, while Democrats tend to support actions that they believe will reduce global warming and its effects through the control of greenhouse gas emissions. A bipartisan measure was introduced in the US House of Representatives as recently as 2007.

The Washington Monument illuminated with a message from Greenpeace criticizing American environmental policy

Climatologist Kevin E. Trenberth stated:

The SPM [Summary for policymakers] was approved line by line by governments[...] The argument here is that the scientists determine what can be said, but the governments determine how it can best be said. Negotiations occur over wording to ensure accuracy, balance, clarity of message, and relevance to understanding and policy. The IPCC process is dependent on the good will of the participants in producing a balanced assessment. However, in Shanghai, it appeared that there were attempts to blunt, and perhaps obfuscate, the messages in the report, most notably by Saudi Arabia. This led to very protracted debates over wording on even bland and what should be uncontroversial text... The most contentious paragraph in the IPCC (2001) SPM was the concluding one on attribution. After much debate, the following was carefully crafted: "In the light of new evidence, and taking into account the remaining uncertainties, most of the observed warming over the last 50 years is likely to have been due to the increase in greenhouse-gas concentrations."

As more evidence has become available over the existence of global warming debate has moved to further controversial issues, including:

1. The social and environmental impacts

2. The appropriate response to climate change

3. Whether decisions require less uncertainty

The single largest issue is the importance of a few degrees rise in temperature:

Most people say, "A few degrees? So what? If I change my thermostat a few degrees, I'll live fine." ... [The] point is that one or two degrees is about the experience that we have had in the last 10,000 years, the era of human civilization. There haven't been—globally averaged, we're talking—fluctuations of more than a degree or so. So we're actually getting into uncharted territory from the point of view of the relatively benign climate of the last 10,000 years, if we warm up more than a degree or two. (Stephen H. Schneider)

The other point that leads to major controversy—because it could have significant economic impacts—is whether action (usually, restrictions on the use of fossil fuels to reduce carbon-dioxide emissions) should be taken now, or in the near future; and whether those restrictions would have any meaningful effect on global temperature.

Because of the economic ramifications of such restrictions, there are those, including the Cato Institute, a libertarian think tank, who argue that the negative economic effects of emission controls outweigh the environmental benefits. They state that even if global warming is caused solely by the burning of fossil fuels, restricting their use would have more damaging effects on the world economy than the increases in global temperature.

The linkage between coal, electricity, and economic growth in the United States is as clear as it can be. And it is required for the way we live, the way we work, for our economic success, and for our future. Coal-fired electricity generation. It is necessary.(Fred Palmer, President of Western Fuels Association)

Conversely, others argue that early action to reduce emissions would help avoid much greater economic costs later, and would reduce the risk of catastrophic, irreversible change. In his December 2006 book, *Hell and High Water*, Joseph J. Romm

discusses the urgency to act and the sad fact that America is refusing to do so...

On a local or regional level, some specific effects of global warming might be considered beneficial.

Council on Foreign Relations senior fellow Walter Russell Mead argues that the 2009 Copenhagen Summit failed because environmentalists have changed from "Bambi to Godzilla". According to Mead, environmentalist used to represent the skeptical few who made valid arguments against big government programs which tried to impose simple but massive solutions on complex situations. Environmentalists' more recent advocacy for big economic and social intervention against global warming, according to Mead, has made them, "the voice of the establishment, of the tenured, of the technocrats" and thus has lost them the support of a public which is increasingly skeptical of global warming.

Various campaigns such as 350.org and many Greenpeace projects have been started in an effort to push the world's leaders towards changing laws and policies that would effectively reduce the world's carbon emissions and use of non-renewable energy resources.

Kyoto Protocol

The Kyoto protocol is the most prominent international agreement on climate change, and is also highly controversial. Some argue that it goes too far or not nearly far enough in restricting emissions of greenhouse gases. Another area of controversy is the fact that China and India, the world's two most populous countries, both ratified the protocol but are not required to reduce or even limit the growth of carbon emissions under the present agreement even though when listed by greenhouse gas emissions per capita, they have rankings of 121st largest per capita emitter at 3.9 Tonnes of CO_2e and 162nd largest per capita emitter at 1.8 Tonnes of CO_2e respectively, compared with for example the US at position of the 14th largest per capita CO_2e emitter at 22.9 Tonnes of CO_2e. Nevertheless, China is the world's second largest producer of greenhouse gas emissions, and India 4th. Various predictions see China overtaking the US in total greenhouse emissions between late 2007 and 2010, and according to many other estimates, this already occurred in 2006.

Additionally, high costs of decreasing emissions may cause significant production to move to countries that are not covered under the treaty, such as India and China, says Fred Singer. As these countries are less energy efficient, this scenario is said to cause additional carbon emissions.

In May 2010 the Hartwell Paper was published by the London School of Economics in collaboration with the University of Oxford. This paper was written by 14 academics from various disciplines in the sciences and humanities, and also some policies thinkers, and they argued that the Kyoto Protocol crashed in late 2009 and "has failed to produce any discernable real world reductions in emissions of greenhouse gases in fifteen years." They argued that this failure opened an opportunity to set climate policy free from Kyoto and the paper advocates a controversial and piecemeal approach to decarbonization of the global economy. The Hartwell paper proposes, "the organising principle of our effort should be the raising up of human dignity via three overarching objectives: ensuring energy access for all; ensuring that we develop in a manner that does not undermine the essential functioning of the Earth system; ensuring that our societies are adequately equipped to withstand the risks and dangers that come from all the vagaries of climate, whatever their cause may be".

The only major developed nation which has signed but not ratified the Kyoto protocol is the US. The countries with no official position on Kyoto are mainly African countries with underdeveloped scientific infrastructure or are oil producers.

Funding

The Global Climate Coalition was an industry coalition that funded several scientists who expressed skepticism about global warming. In the year 2000, several members left the coalition when they became the target of a national divestiture campaign run by John Passacantando and Phil Radford at Ozone Action. According to *The New York Times*, when Ford Motor Company was the first company to leave the coalition, it was "the latest sign of divisions within heavy industry over how to respond to global warming". After that, between December, 1999 and early March, 2000, the GCC was deserted by Daimler-Chrysler, Texaco, energy firm the Southern Company and General Motors. The Global Climate Coalition closed in 2002, or in their own words, 'deactivated'.

Documents obtained by Greenpeace under the US Freedom of Information Act show that the Charles G. Koch Foundation gave climate change writer Willie Soon two grants totaling $175,000 in 2005/6 and again in 2010. Multiple grants to Soon from the American Petroleum Institute between 2001 and 2007 totalled $274,000, and from ExxonMobil totalled $335,000 between 2005 and 2010. Other coal and oil industry sources which funded him include the Mobil Foundation, the Texaco Foundation and the Electric Power Research Institute. Soon, acknowledging that he received this money, stated unequivocally that he has "never been motivated by financial reward in any of my scientific research". In February 2015, Greenpeace disclosed papers documenting that Soon failed to disclose to academic journals funding including more than $1.2 million from fossil fuel industry related interests including ExxonMobil, the American Petroleum Institute, the Charles G. Koch Charitable Foundation and the Southern Company. To investigate how widespread such hidden funding was, senators Barbara Boxer, Edward Markey and Sheldon Whitehouse wrote to a number of companies. Koch general counsel refused the request and said it would infringe the company's first amendment rights.

The Greenpeace research project ExxonSecrets, and George Monbiot writing in *The Guardian*, as well as various academics, have linked several skeptical scientists—Fred Singer, Fred Seitz and Patrick Michaels—to organizations funded by ExxonMobil and Philip Morris for the purpose of promoting global warming skepticism. These organizations include the Cato Institute and the Heritage Foundation. Similarly, groups employing global warming skeptics, such as the George C. Marshall Institute, have been criticized for their ties to fossil fuel companies.

On 2 February 2007, *The Guardian* stated that Kenneth Green, a Visiting Scholar with AEI, had sent letters to scientists in the UK and the U.S., offering US$10,000 plus travel expenses and other incidental payments in return for essays with the purpose of "highlight[ing] the strengths and weaknesses of the IPCC process", specifically regarding the IPCC Fourth Assessment Report.

A furor was raised when it was revealed that the Intermountain Rural Electric Association (an energy cooperative that draws a significant portion of its electricity from coal-burning power plants) donated $100,000 to Patrick Michaels and his group, New Hope Environmental Services, and solicited additional private donations from its members.

The Union of Concerned Scientists produced a report titled 'Smoke, Mirrors & Hot Air', that criticizes ExxonMobil for "underwriting the most sophisticated and most successful disinformation campaign since the tobacco industry" and for "funnelling about $16 million between 1998 and 2005 to a network of ideological and advocacy organizations that manufacture uncertainty on the issue". In 2006 Exxon said that it was no longer going to fund these groups though that statement has been challenged by Greenpeace.

The Center for the Study of Carbon Dioxide and Global Change, a skeptic group, when confronted about the funding of a video they put together ($250,000 for "The Greening of Planet Earth" from an oil company) stated, "We applaud Western Fuels for their willingness to publicize a side of the story that we believe to be far more correct than what at one time was 'generally accepted'. But does this mean that they fund The Center? Maybe it means that we fund them!"

Donald Kennedy, editor-in-chief of *Science*, has said that skeptics such as Michaels are lobbyists more than researchers, and "I don't think it's unethical any more than most lobbying is unethical," he said. He said donations to skeptics amounts to "trying to get a political message across".

Global warming skeptic Reid Bryson said in June 2007, "There is a lot of money to be made in this... If you want to be an eminent scientist you have to have a lot of grad students and a lot of grants. You can't get grants unless you say, 'Oh global warming, yes, yes, carbon dioxide'." Similar positions have been advanced by University of Alabama, Huntsville climate scientist Roy Spencer, Spencer's University of Alabama, Huntsville colleague and IPCC contributor John Christy, University of London biogeographer Philip Stott, Accuracy in Media, and Ian Plimer in his 2009 book *Heaven and Earth — Global Warming: The Missing Science.*

Richard Lindzen, the Alfred P. Sloan Professor of Meteorology at MIT, said, "[in] the winter of 1989 Reginald Newell, a professor of meteorology [at MIT], lost National Science Foundation funding for data analyses that were failing to show net warming over the past century." Lindzen also suggested that four other scientists "apparently" lost their funding or positions after questioning the scientific underpinnings of global warming. Lindzen himself has been the recipient of money from energy interests such as OPEC and the Western Fuels Association, including "$2,500 a day for his consulting services", as well as funding from US federal sources including the National Science Foundation, the Department of Energy, and NASA.

Debate Over Most Effective Response to Warming

In recent years some skeptics have changed their positions regarding global warming. Ronald Bailey, author of *Global Warming and Other Eco-Myths* (published by the Competitive Enterprise Institute in 2002), stated in 2005, "Anyone still holding onto the idea that there is no global warming ought to hang it up." By 2007, he wrote "Details like sea level rise will continue to be debated by researchers, but if the debate over whether or not humanity is contributing to global warming wasn't over before, it is now.... as the new IPCC Summary makes clear, climate change Pollyannaism is no longer looking very tenable."

"There are alternatives to its [the climate-change crusade's] insistence that the only appropriate policy response is steep and immediate emissions reductions.... a greenhouse-gas-emissions cap ultimately would constrain energy production. A sensible climate policy would emphasize building resilience into our capacity to adapt to climate changes.... we should consider strategies of adaptation to a changing climate. A rise in the sea level need not be the end of the world, as the Dutch have taught us." says Steven F. Hayward of American Enterprise Institute, a conservative think-tank. Hayward also advocates the use of "orbiting mirrors to rebalance the amounts of solar radiation different parts of the earth receive"—the space sunshade example of so-called geoengineering for solar radiation management.

In 2001 Richard Lindzen, asked whether it was necessary to try to reduce CO_2 emissions, said that responses needed to be prioritised. "You can't just say, 'No matter what the cost, and no matter how little the benefit, we'll do this'. If we truly believe in warming, then we've already decided we're going to adjust...The reason we adjust to things far better than Bangladesh is that we're richer. Wouldn't you think it makes sense to make sure we're as robust and wealthy as possible? And that the poor of the world are also as robust and wealthy as possible?"

Others argue that if developing nations reach the wealth level of the United States this could greatly increase CO_2 emissions and consumption of fossil fuels. Large developing nations such as India and China are predicted to be major emitters of greenhouse gases in the next few decades as their economies grow.

The conservative National Center for Policy Analysis whose "Environmental Task Force" contains a number of climate change skeptics including Sherwood Idso and S. Fred Singer says, "The growing consensus on climate change policies is that adaptation will protect present and future generations from climate-sensitive risks far more than efforts to restrict CO_2 emissions."

The adaptation-only plan is also endorsed by oil companies like ExxonMobil, "ExxonMobil's plan appears to be to stay the course and try to adjust when changes occur. The company's plan is one that involves adaptation, as opposed to leadership," says this Ceres report.

Gregg Easterbrook characterized himself as having "a long record of opposing alarmism". In 2006, he stated, "based on the data I'm now switching sides regarding global warming, from skeptic to convert."

The George W. Bush administration also voiced support for an adaptation-only policy in the US in 2002. "In a stark shift for the Bush administration, the United States has sent a climate report [*U.S. Climate Action Report 2002*] to the United Nations detailing specific and far-reaching effects it says global warming will inflict on the American environment. In the report, the administration also for the first time places most of the blame for recent global warming on human actions—mainly the burning of fossil fuels that send heat-trapping greenhouse gases into the atmosphere." The report however "does not propose any major shift in the administration's policy on greenhouse gases. Instead it recommends adapting to inevitable changes instead of making rapid and drastic reductions in greenhouse gases to limit warming." This position apparently precipitated a similar shift in emphasis at the COP 8 climate talks in New Delhi several months later, "The shift satisfies the Bush administration, which has fought to avoid mandatory cuts in emissions for fear it would harm the economy. 'We're welcoming a focus on more of a balance on adaptation versus mitigation', said a senior American negotiator in New Delhi. 'You don't have enough money to do everything.'" The White House emphasis on adaptation was not well received however:

Despite conceding that our consumption of fossil fuels is causing serious damage and despite implying that current policy is inadequate, the Report fails to take the next step and recommend serious alternatives. Rather, it suggests that we simply need to accommodate to the coming changes. For example, reminiscent of former Interior Secretary Hodel's proposal that the government address the hole in the ozone layer by encouraging Americans to make better use of sunglasses, suntan lotion and broad-brimmed hats, the Report suggests that we can deal with heat-related health impacts by increased use of air-conditioning ... Far from proposing solutions to the climate change problem, the Administration has been adopting energy policies that would actually increase greenhouse gas emissions. Notably, even as the Report identifies increased air conditioner use as one of the 'solutions' to climate change impacts, the Department of Energy has decided to roll back energy efficiency standards for air conditioners.

— Letter from 11 State Attorneys General to George W. Bush.,

Some find this shift and attitude disingenuous and indicative of an inherent bias against prevention (i.e. reducing emissions/consumption) and for the prolonging of profits to the oil industry at the expense of the environment. "Now that the dismissal of climate change is no longer fashionable, the professional deniers are trying another means of stopping us from taking action. It would be cheaper, they say, to wait for the impacts of climate change and then adapt to them" says writer and environmental activist George Monbiot in an article addressing the supposed economic hazards of addressing climate change. Others argue that adaptation alone will not be sufficient.

Though not emphasized to the same degree as mitigation, adaptation to a climate certain to change has been included as a necessary component in the discussion as early as 1992, and has been all along. However it was not to the *exclusion*, advocated by the skeptics, of *preventative* mitigation efforts, and therein, say carbon cutting proponents, lies the difference.

Another highly debated potential climate change mitigation strategy is Cap and Trade due to its direct relationship with the economy.

Political Pressure on Scientists

Many climate scientists state that they are put under enormous pressure to distort or hide any scientific results which suggest that human activity is to blame for global warming. A survey of climate scientists which was reported to the US House Oversight and Government Reform Committee in 2007 noted "Nearly half of all respondents perceived or personally experienced pressure to eliminate the words 'climate change', 'global warming' or other similar terms from a variety of communications." These scientists were pressured to tailor their reports on global warming to fit the Bush administration's climate change scepticism. In some cases, this occurred at the request of former oil-industry lobbyist Phil Cooney, who worked for the American Petroleum Institute before becoming chief of staff at the White House Council on Environmental Quality (he resigned in 2005 before being hired by ExxonMobil). In June 2008, a report by NASA's Office of the Inspector General concluded that NASA staff appointed by the White House had censored and suppressed scientific data on global warming in order to protect the Bush administration from controversy close to the 2004 presidential election.

U.S. officials, such as Philip Cooney, have repeatedly edited scientific reports from US government scientists, many of whom, such as Thomas Knutson, have been ordered to refrain from discussing climate change and related topics. Attempts to suppress scientific information on global warming and other issues have been described by journalist Chris Mooney in his book *The Republican War on Science.*

Climate scientist James E. Hansen, director of NASA's Goddard Institute for Space Studies, wrote in a widely cited *New York Times* article in 2006 that his superiors at the agency were trying to "censor" information "going out to the public". NASA denied this, saying that it was merely requiring that scientists make a distinction between personal, and official government, views in interviews conducted as part of work done at the agency. Several scientists working at the National Oceanic and Atmospheric Administration have made similar complaints; once again, government officials said they were enforcing long-standing policies requiring government scientists to clearly identify personal opinions as such when participating in public interviews and forums.

The BBC's long-running current affairs series *Panorama* recently investigated the issue, and was told, "scientific reports about global warming have been systematically changed and suppressed."

Scientists who agree with the consensus view have sometimes expressed concerns over what they view as sensationalism of global warming by interest groups and the press. For example, Mike Hulme, director of the Tyndall Centre for Climate Research, wrote how increasing use of pejorative terms like "catastrophic", "chaotic" and "irreversible", had altered the public discourse around climate change: "This discourse is now characterised by phrases such as 'climate change is worse than we thought', that we are approaching 'irreversible tipping in the Earth's climate', and that we are 'at the point of no return'. I have found myself increasingly chastised by climate change campaigners when my public statements and lectures on climate change have not satisfied their thirst for environmental drama and exaggerated rhetoric."

According to an Associated Press release on 30 January 2007,

Climate scientists at seven government agencies say they have been subjected to political pressure aimed at downplaying the threat of global warming.

The groups presented a survey that shows two in five of the 279 climate scientists who responded to a questionnaire complained that some of their scientific papers had been edited in a way that changed their meaning. Nearly half of the 279 said in response to another question that at some point they had been told to delete reference to "global warming" or "climate change" from a report.

Critics writing in *The Wall Street Journal* editorial page state that the survey was itself unscientific.

In addition to the pressure from politicians, many prominent scientists working on climate change issues have reported increasingly severe harassment from members of the public. The harassment has taken several forms. The US FBI told ABC News that it was looking into a spike in threatening emails sent to climate scientists, while a white supremacist website posted pictures of several climate scientists with the word "Jew" next to each image. One climate scientist interviewed by ABC News had a dead animal dumped on his doorstep and now frequently has to travel with bodyguards.

In April 2010, Virginia Attorney General Ken Cuccinelli claimed that leading climate scientist Michael E. Mann had possibly violated state fraud laws, and without providing any evidence of wrongdoing, filed the Attorney General of Virginia's climate science investigation as a civil demand that the University of Virginia provide a wide range of records broadly related to five research grants Mann had obtained as an assistant professor at the university from 1999 to 2005. This litigation was widely criticized in the academic community as politically motivated and likely to have a chilling effect on future research. The university filed a court petition and the judge dismissed Cuccinelli's demand on the grounds that no justification had been shown for the investigation. Cuccinelli issued a revised subpoena, and appealed the case to the Virginia Supreme Court which ruled in March 2012 that Cuccinelli did not have the authority to make these demands. The outcome was hailed as a victory for academic freedom.

Exxon Mobil is also notorious for skewing scientific evidence through their private funding of scientific organizations. In 2002, Exxon Mobil contributed $10,000 to The Independent Institute and then $10,000 more in 2003. In 2003, The Independent Institute release a study that reported the evidence for imminent global warming found during the Clinton administration was based on now-dated satellite findings and wrote off the evidence and findings as a product of "bad science".

This is not the only consortium of skeptics that Exxon Mobil has supported financially. The George C. Marshall Institute received $630,000 in funding for climate change research from ExxonMobil between 1998 and 2005. Exxon Mobil also gave $472,000 in funding to The Board of Academic and Scientific Advisors for the Committee for a Constructive Tomorrow from 1998 to 2005. Dr. Frederick Seitz, well known as "the godfather of global warming skepticism", served as both Chairman Emeritus of The George C. Marshall Institute and a board member of the Committee for a Constructive Tomorrow from 1998 to 2005.

Litigation

Several lawsuits have been filed over global warming. For example, Massachusetts v. Environmental Protection Agency before the Supreme Court of the United States allowed the EPA to regulate greenhouse gases under the Clean Air Act. A similar approach was taken by California Attorney General Bill Lockyer who filed a lawsuit California v. General Motors Corp. to force car manufacturers to reduce vehicles' emissions of carbon dioxide. This lawsuit was found to lack legal merit and was tossed out. A third case, Comer v. Murphy Oil USA, Inc., a class action lawsuit filed by Gerald Maples, a trial attorney in Mississippi, in an effort to force fossil fuel and chemical companies to pay for damages caused by global warming. Described as a nuisance lawsuit, it was dismissed by District Court. However, the District Court's decision was overturned by the United States Court of Appeals for the Fifth Circuit, which instructed the District Court to reinstate several of the plaintiffs' climate change-related claims on 22 October 2009. The Sierra Club sued the U.S. government over failure to raise automobile fuel efficiency standards, and thereby decrease carbon dioxide emissions.

Kelsey Cascade, Rose Juliana et. al. vs. United States

In a lawsuit organized by activist organization Our Children's Trust, a group of plaintiffs aged 8–19 sued the U. S. Federal Government, claiming "the government has known for decades that carbon dioxide (CO2) pollution has been causing catastrophic climate change and has failed to take necessary action to curtail fossil fuel emissions." On April 8, 2016, U. S. Magistrate Judge Thomas Coffin denied defendant's motion to dismiss, arguing plaintiffs have standing to sue because they will be disproportionately affected by the alleged damages. "The intractability of the debates before Congress and state legislatures and the alleged valuing of short term economic interest despite the cost to human life," argued Coffin, "necessitates a need for the courts to evaluate the constitutional parameters of the action or inaction taken by the government".

Extreme Weather

Extreme weather includes unexpectable, unusual, unpredictable severe or unseasonal weather; weather at the extremes of the historical distribution—the range that has been seen in the past. Often, extreme events are based on a location's recorded weather history and defined as lying in the most unusual ten percent. In recent years some extreme weather events have been attributed to human-induced global warming, with studies indicating an increasing threat from extreme weather in the future.

A tornado that struck Anadarko, Oklahoma during a tornado outbreak in 1999

Costs

According to IPCC (2011) estimates of annual losses have ranged since 1980 from a few billion to above US$200 billion (in 2010 dollars), with the highest value for 2005 (the year of Hurricane Katrina) . The global weather-related disaster losses because many impacts, such as loss of human lives, cultural heritage, and ecosystem services, are difficult to value and monetize, and thus they are poorly reflected in estimates of losses.

Extreme Temperatures

Heat Waves

Heat waves are periods of abnormally high temperatures and heat index. Definitions of a heatwave vary because of the variation of temperatures in different geographic locations. Excessive heat is often accompanied by high levels of humidity, but can also be catastrophically dry.

Because heatwaves are not visible as other forms of severe weather are, like hurricanes, tornadoes, and thunderstorms, they are one of the less known forms of extreme weather. Severe heat weather can damage populations and crops due to potential dehydration or hyperthermia, heat cramps, heat expansion and heat stroke. Dried soils are more susceptible to erosion, decreasing lands available for agriculture. Outbreaks of wildfires can increase in frequency as dry vegetation has increased likeliness of igniting. The evaporation of bodies of water can be devastating to marine populations, decreasing the size of the habitats available as well as the amount of nutrition present within the waters. Livestock and other animal populations may decline as well.

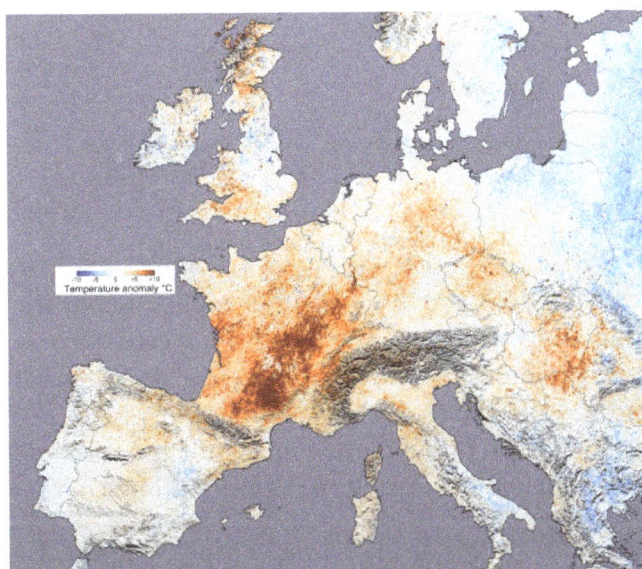

2003 European heat wave

During excessive heat plants shut their leaf pores (stomata), a protective mechanism to conserve water but also curtails plants' absorption capabilities. Thus, leaving more pollution and ozone in the air, which leads to a higher mortality in the population. It has been estimated that extra pollution during the hot summer 2006 in the UK, cost 460 lives. The European heat waves from summer 2003 are estimated to have caused 30,000 excess deaths, due to heat stress and air pollution.

Power outages can also occur within areas experiencing heat waves due to the increased demand for electricity (i.e. air conditioning use). The urban heat island effect can increase temperatures, particularly overnight.

Cold Waves

A cold wave is a weather phenomenon that is distinguished by a cooling of the air. Specifically, as used by the U.S. National Weather Service, a cold wave is a rapid fall in temperature within a 24-hour period requiring substantially increased protection to agriculture, industry, commerce, and social activities. The precise criterion for a cold wave is determined by the rate at which the temperature falls, and the minimum to which it falls. This minimum temperature is dependent on the geographical region and time of year. Cold waves generally are capable of occurring any geological location and are formed by large cool air masses that accumulate over certain regions, caused by movements of air streams.

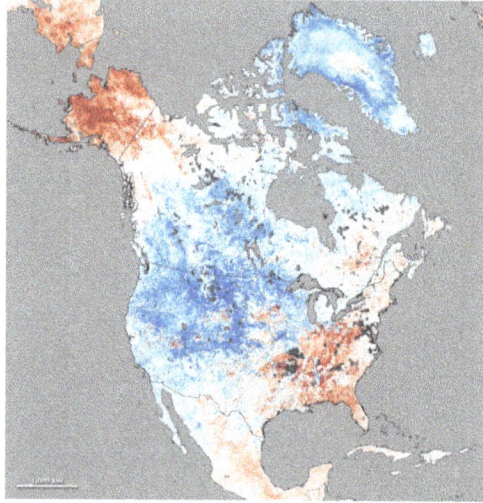

Cold wave in continental North America since Dec-03 to Dec-10, 2013. Red color means above mean temperature; blue represents below normal temperature.

A cold wave can cause death and injury to livestock and wildlife. Exposure to cold mandates greater caloric intake for all animals, including humans, and if a cold wave is accompanied by heavy and persistent snow, grazing animals may be unable to reach necessary food and water, and die of hypothermia or starvation. Cold waves often necessitate the purchase of fodder for livestock at considerable cost to farmers. Human populations can be inflicted with frostbites when exposed for extended periods of time to cold and may result in the loss of limbs or damage to internal organs.

Extreme winter cold often causes poorly insulated water pipes to freeze. Even some poorly protected indoor plumbing may rupture as frozen water expands within them, causing property damage. Fires, paradoxically, become more hazardous during extreme cold. Water mains may break and water supplies may become unreliable, making firefighting more difficult.

Cold waves that bring unexpected freezes and frosts during the growing season in mid-latitude zones can kill plants during the early and most vulnerable stages of growth. This results in crop failure as plants are killed before they can be harvested economically. Such cold waves have caused famines. Cold waves can also cause soil particles to harden and freeze, making it harder for plants and vegetation to grow within these areas. One extreme was the so-called Year Without a Summer of 1816, one of several years during the 1810s in which numerous crops failed during freakish summer cold snaps after volcanic eruptions reduced incoming sunlight.

Climate Change

In general climate models show that with climate change, the planet will experience more extreme weather. In particular temperature record highs outpace record lows and some types of extreme weather such as extreme heat, intense precipitation, and drought have become more frequent and severe in recent decades. Some studies assert a connection between rapidly warming arctic temperatures and thus a vanishing cryosphere to extreme weather in mid-latitudes.

Heat Stress

The upper limit for heat stress humans can adapt to is called into question with a 7 °C temperature rise, quantified by the wet-bulb temperature, regions of Earth would lose their habitability.

Tropical Cyclones

There has been long ongoing debate about a possible increase of tropical cyclones as an effect of global warming. However, the 2012 IPCC special report on extreme events SREX states that "there is low confidence in any observed long-term (i.e., 40 years or more) increases in tropical cyclone activity (i.e., intensity, frequency, duration), after accounting for past changes in observing capabilities." Increases in population densities increase the number of people affected and damage caused by an event of given severity. The World Meteorological Organization and the U.S. Environmental Protection Agency have in the past linked increasing extreme weather events to global warming, as have Hoyos *et al.* (2006), writing that the increasing number of category 4 and 5 hurricanes is directly linked to increasing temperatures. Similarly, Kerry Emanuel in *Nature* writes that hurricane power dissipation is highly correlated with temperature, reflecting global warming.

Hurricane modeling has produced similar results, finding that hurricanes, simulated under warmer, high CO_2 conditions, are more intense than under present-day conditions. Thomas Knutson and Robert E. Tuleya of the NOAA stated in 2004 that warming induced by greenhouse gas may lead to increasing occurrence of highly destructive category-5 storms. Vecchi and Soden find that wind shear, the increase of which acts to inhibit tropical cyclones, also changes in model-projections of global warming. There are projected increases of wind shear in the tropical Atlantic and East Pacific associated with the deceleration of the Walker circulation, as well as decreases of wind shear in the western and central Pacific. The study does not make claims about the net effect on Atlantic and East Pacific hurricanes of the warming and moistening atmospheres, and the model-projected increases in Atlantic wind shear.

References

- IPCC AR4 SYR (2007). Core Writing Team; Pachauri, R.K; Reisinger, A., eds. Climate Change 2007: Synthesis Report. Contribution of Working Groups I, II and III to the Fourth Assessment Report of the Intergovernmental Panel on Climate Change. IPCC. ISBN 92-9169-122-4.

- IPCC AR4 WG1 (2007). Solomon, S.; Qin, D.; Manning, M.; Chen, Z.; Marquis, M.; Averyt, K.B.; Tignor, M.; Miller, H.L., eds. Climate Change 2007: The Physical Science Basis. Contribution of Working Group I to the Fourth Assessment Report of the Intergovernmental Panel on Climate Change. Cambridge University Press. ISBN 978-0-521-88009-1.

- IPCC AR4 WG2 (2007). Parry, M.L.; Canziani, O.F.; Palutikof, J.P.; van der Linden, P.J.; Hanson, C.E., eds. Climate Change 2007: Impacts, Adaptation and Vulnerability. Contribution of Working Group II to the Fourth Assessment Report of the Intergovernmental Panel on Climate Change. Cambridge University Press. ISBN 978-0-521-88010-7.

- IPCC AR4 WG3 (2007). Metz, B.; Davidson, O.R.; Bosch, P.R.; Dave, R.; Meyer, L.A., eds. Climate Change 2007: Mitigation of Climate Change. Contribution of Working Group III to the Fourth Assessment Report of the Intergovernmental Panel on Climate Change. Cambridge University Press. ISBN 978-0-521-88011-4.

- IPCC SAR WG3 (1996). Bruce, J.P.; Lee, H.; Haites, E.F., eds. Climate Change 1995: Economic and Social Dimensions of Climate Change. Contribution of Working Group III to the Second Assessment Report of the

Intergovernmental Panel on Climate Change. Cambridge University Press. ISBN 0-521-56051-9.

- IPCC TAR WG1 (2001). Houghton, J.T.; Ding, Y.; Griggs, D.J.; Noguer, M.; van der Linden, P.J.; Dai, X.; Maskell, K.; Johnson, C.A., eds. Climate Change 2001: The Scientific Basis. Contribution of Working Group I to the Third Assessment Report of the Intergovernmental Panel on Climate Change. Cambridge University Press. ISBN 0-521-80767-0.

- IPCC TAR WG3 (2001). Metz, B.; Davidson, O.; Swart, R.; Pan, J., eds. Climate Change 2001: Mitigation. Contribution of Working Group III to the Third Assessment Report of the Intergovernmental Panel on Climate Change. Cambridge University Press. ISBN 0-521-80769-7. (pb: 0-521-01502-2)

- National Research Council (2010). America's Climate Choices: Panel on Advancing the Science of Climate Change;. Washington, D.C.: The National Academies Press. ISBN 0-309-14588-0.

- Doan, Lynn; Covarrubias, Amanda (2006-07-27). "Heat Eases, but Thousands of Southern Californians Still Lack Power". Los Angeles Times. Retrieved June 16, 2014.

- Redfern, Simon (November 8, 2013). "Super Typhoon Haiyan hits Philippines with devastating force". Theconversation.com. Retrieved 2014-08-25.

- "Geophysical Fluid Dynamics Laboratory - Global Warming and 21st Century Hurricanes". Gfdl.noaa.gov. 2014-08-04. Retrieved 2014-08-25.

Permissions

Index